Computing Systems for Autonomous Driving

Weisong Shi • Liangkai Liu

Computing Systems for Autonomous Driving

 Springer

Weisong Shi
Department of Computer Science
Wayne State University
Detroit, MI, USA

Liangkai Liu
Department of Computer Science
Wayne State University
Detroit, MI, USA

ISBN 978-3-030-81566-0   ISBN 978-3-030-81564-6   (eBook)
https://doi.org/10.1007/978-3-030-81564-6

This Springer imprint is published by the registered company Springer Nature Switzerland AG
The registered company address is: Gewerbestrasse 11, 6330 Cham, Switzerland

Preface

In the last 5 years, with the vast improvements in computing technologies, e.g., sensors, computer vision, machine learning, and hardware acceleration, and the wide deployment of communication mechanisms, e.g., dedicated short-range communications (DSRC), cellular vehicle-to-everything (C-V2X), and 5G, autonomous driving techniques have attracted massive attention from both the academic and automotive communities.

To achieve the vision of autonomous driving, determining how to make the vehicle understand the environment correctly and make safe controls in real-time is the essential task. Rich sensors including camera, LiDAR (light detection and ranging), radar, inertial measurement unit (IMU), global navigation satellite system (GNSS), and sonar, as well as powerful computation devices, are installed on the vehicle. This design makes autonomous driving a real powerful "computer on wheels." In addition to hardware, the rapid development of deep learning algorithms in object/lane detection, simultaneous localization and mapping (SLAM), and vehicle control also promotes the real deployment and prototyping of autonomous vehicles. The autonomous vehicle's computing systems are defined to cover everything (excluding the vehicle's mechanical parts), including sensors, computation, communication, storage, power management, and full-stack software. Plenty of algorithms and systems are designed to process sensor data and make a reliable decision in real-time.

However, news of fatalities caused by early developed autonomous vehicles (AVs) arises from time to time. Until August 2020, five self-driving car fatalities happened for level 2 autonomous driving: four of them from Tesla and one from Uber. All four incidents associated with Tesla are due to perception failure, while Uber's incident happened because of the failure to predict human behavior. Another fact to pay attention to is that currently, the field-testing of level 2 autonomous driving vehicles mostly happens in places with good weather and light traffic conditions like Arizona and Florida. The real traffic environment is too complicated for the current autonomous driving systems to understand and handle easily. The objectives of level 4 and level 5 autonomous driving require colossal improvement of the computing systems for autonomous vehicles.

This book intends to present state-of-the-art computing systems for autonomous driving and to grab the attention of researchers and practitioners from both automotive industry and computer science and engineering community. The book consists of nine chapters, presenting the landscape, computing frameworks, algorithm deployment optimizations, systems runtime optimizations, dataset and benchmarking, simulators, hardware platforms, smart infrastructures, and open challenges for achieving L4/L5 autonomous driving vehicles, respectively. This book can be used by senior undergraduate students and graduate students in engineering and computer science majors. We hope this book will serve as a reference and a starting point for those who are interested in working in this field.

Detroit, MI, USA   Weisong Shi

Detroit, MI, USA   Liangkai Liu

Acknowledgments

This book is a collective wisdom of the work from the Connected and Autonomous Driving Laboratory (CAR) at Wayne State University. We would like to thank all the past and current members in the CAR lab, including Sidi Lu, Qingyang Zhang, Yifan Wang, Xingzhou Zhang, Baofu Wu, Prabhjot Kaur, Samira Taghavi, Ren Zhong, Yongtao Yao, Ruijun Wang, Zhaofeng Tian, and Raef Abdallah. All of them contributed part of the content that is included in this book. We also thank our partners and sponsors who contributed hardware, software, and dataset and made these studies possible, including CalmCar, Continental, DENSO, Hesai, iSmartWays, Intel, Navya, Nvidia, PerceptIn, Toyota InfoTech, Velodyne LiDAR, Xilinx, and the City of Detroit.

Contents

Chapter 1
Autonomous Driving Landscape

1.1 Reference Architecture

As an essential part of the whole autonomous driving vehicle, the computing system plays a significant role in the whole pipeline of driving autonomously. There are two types of designs for computing systems on autonomous vehicles: modular-based and end-to-end based.

Modular design decouples the localization, perception, control, etc. as separate modules and makes it possible for people with different backgrounds to work together [254]. The DARPA challenges are a milestone for the prototyping of autonomous driving vehicles, including Boss from CMU [531], Junior from Stanford [292], TerraMax and BRAiVE from University of Parma [61], etc. Their designs are all based on modules including perception, mission planning, motion planning, and vehicle controls. Similarly, the survey fleet vehicles developed by Google and Uber are also modular-based [49, 490]. The main differences for these AV prototypes are the software and the configuration of sensors like camera, LiDAR, Radar, etc.

In contrast, the end-to-end based design is largely motivated by the development of artificial intelligence. Compared with modular design, end-to-end system purely relies on machine learning techniques to process the sensor data and generate control commands to the vehicle [52, 75, 257, 368, 456, 571]. Table 1.1 shows a detailed description of these end-to-end designs. Four of them are based on supervised DNNs to learn driving patterns and behaviors from human drivers. The remaining two are based on Deep Q-Network (DQN), which learns to find the optimum driving by itself. Although the end-to-end based approach promises to decrease the modular design's error propagation and computation complexity, there is no real deployment and testing of it [589].

As most prototypes are still modular-based, we choose it as the basis for the computing system reference architecture. Figure 1.1 shows a representative reference architecture of the computing system on autonomous vehicles. Generally,

© The Author(s), under exclusive license to Springer Nature Switzerland AG 2021
W. Shi, L. Liu, *Computing Systems for Autonomous Driving*,
https://doi.org/10.1007/978-3-030-81564-6_1

Table 1.1 End-to-end approaches for autonomous driving

Work	Methods	Characteristics
[368]	Supervised DNN	Raw image to steering angles for off-road obstacle avoidance on mobile robots
[75]	Supervised DNN	Map an input image to a small number of key perception indicators
[52]	Supervised DNN	CNN to map raw pixels from a camera directly to steering commands
[571]	Supervised DNN	FCN-LSTM network to predict multi-modal discrete and continuous driving behaviors
[456]	DQN	Automated driving framework in simulator environment
[257]	DQN	Lane following in a countryside road without traffic using a monocular image as input

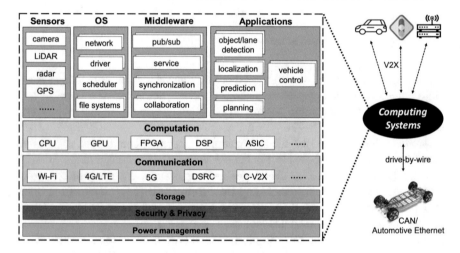

Fig. 1.1 Representative reference architecture of the computing system for autonomous driving

the computing system for autonomous driving vehicles can be divided into computation, communication, storage, security and privacy, and power management. Each part covers four layers with sensors, operating system (OS), middleware, and applications. The following paragraphs will discuss the corresponding components.

For safety, one of the essential tasks is to enable the "computer" to understand the road environment and send correct control messages to the vehicle. The whole pipeline starts with the sensors. Plenty of sensors can be found on an autonomous driving vehicle: camera, LiDAR, radar, GPS/GNSS, ultrasonic, inertial measurement unit (IMU), etc. These sensors capture real-time environment information for the computing system, like the eyes of human beings. Operating system (OS) plays a vital role between hardware devices (sensors, computation, communication) and applications. Within the OS, drivers are bridges between the software and hardware devices; the network module provides the abstraction communication interface; the scheduler manages the competition to all the resources; the file system provides the abstraction to all the resources. For safety-critical scenarios, the operating system must satisfy real-time requirements.

As the middle layer between applications and operating systems [469], middleware provides usability and programmability to develop and improve systems more effectively. Generally, communication middleware supports publisher/subcriber messaging, remote procedure call (RPC) or service, time synchronization, and multi-sensor collaboration. A typical example of the middleware system is the Robot Operating System (ROS) [424]. On top of the operating system and middleware system, several applications, including object/lane detection, SLAM, prediction, planning, and vehicle control, are implemented to generate control commands and send them to the vehicle's drive-by-wire system. Inside the vehicle, several Electronic Control Units (ECUs) are used to control the brake, steering, etc., which are connected via Controller Area Network (CAN bus) or Automotive Ethernet [195]. In addition to processing the data from on-board sensors, the autonomous driving vehicle is also supposed to communicate with other vehicles, traffic infrastructures, pedestrians, etc. as complementary.

1.2   Metrics for Computing System

According to the report about autonomous driving technology from the National Science & Technology Council (NSTC) and the United States Department of Transportation (USDOT) [130], ten technology principles are designed to foster research, development, and integration of AVs and guide consistent policy across the U.S. Government. These principles cover safety, security, cyber security, privacy, data security, mobility, accessibility, etc. Corresponding to the autonomous driving principles, we define several metrics to evaluate the computing system's effectiveness.

Accuracy   Accuracy is defined to evaluate the difference between the detected/processed results with the ground truth. Take object detection and lane detection, for example, the Intersection Over Union (IOU) and mean Average Precision (mAP) are used to calculate the exact difference between the detected bounding box of objects/lanes and the real positions [133, 172]. For vehicle controls, the accuracy would be the difference between the expected controls in break/steering with the vehicle's real controls.

Timeliness   Safety is always the highest priority. Autonomous driving vehicles should be able to control themselves autonomously in real-time. According to [254], if the vehicle is self-driving at 40 km per hour in an urban area and wants the control effective every 1 m, then the whole pipeline's desired response time should be less than 90 ms. To satisfy the desired response time, we need each module in the computing system to finish before the deadline.

Power Since the on-board battery powers the whole computing system, the computing system's power dissipation can be a big issue. For electrical vehicles, the computing system's power dissipation for autonomous driving reduces the vehicle's mileage with up to 30% [25]. In addition to mileage, heat dissipation is another issue caused by high power usage. Currently, the NVIDIA Drive PX Pegasus provides 320 INT8 TOPS of AI computational power with a 500 W budget [350]. With the power budget of sensors, communication devices, etc., the total power dissipation will be higher than 1000 W. The power budget is supposed to be a significant obstacle for producing the real autonomous driving vehicle.

Cost Cost is one of the essential factors that affect the board deployment of autonomous vehicles. According to [177, 494], the cost of a level 4 autonomous driving vehicle attains 300,000 dollars, in which the sensors, computing device, and communication device cost almost 200,000 dollars. In addition to the hardware cost, the operator training and vehicle maintenance cost of AVs (like insurance, parking, and repair) is also more expensive than traditional vehicles.

Reliability To guarantee the safety of the vehicle, reliability is a big concern [453]. On the one hand, the worst-case execution time is supposed to be longer than the deadline. Interruptions or emergency stops should be applied in such cases. On the other hand, failures happen in sensors, computing/communication devices, algorithms, and systems integration [397]. How to handle these potential failures is also an essential part of the design of the computing system.

Privacy As the vehicle captures a massive amount of sensor data from the environment, vehicle data privacy becomes a big issue. For example, the pedestrian's face and the license plate captured by the vehicle's camera should be masked as soon as possible [502]. Furthermore, who owns the driving data is also an important issue, which requires the system's support for data access, storage, and communication [440].

Security The secureness of the on-board computing system is essential to the success of autonomous driving since, ultimately, the computing system is responsible for the driving process. Cyber attacks can be launched quickly to any part of the computing system [440, 580]. We divide the security into four aspects: sensing security, communication security, data security, and control security [574, 592]. We envision that the on-board computing system will have to pass a certain security test level before deploying it into real products.

1.3   Key Technologies

In this section, we summarize several key technologies and discuss their state of the art.

1.3.1   Sensors

Cameras In terms of usability and cost, cameras are the most popular sensors on autonomous driving vehicles. The camera image gives straightforward 2D information, making it useful in some tasks like object classification and lane tracking. Also, the range of the camera can vary from several centimeters to near one hundred meters. The relatively low cost and commercialization production also contribute to the complete deployment in the real autonomous driving vehicle. However, based on lights, the camera's image can be affected by low lighting or bad weather conditions. The usability of the camera decreases significantly under heavy fog, raining, and snowing. Furthermore, the data from the camera is also a big problem. On average, every second, one camera can produce 20–40 MB of data.

Radar The radar's full name is Radio Detection and Ranging, which means to detect and get the distance using radio. The radar technique measures the Time of Flight (TOF) and calculates the distance and speed. Generally, the working frequency of the vehicle radar system is 24 GHz or 77 GHz. Compared with 24 GHz, 77 GHz shows higher accuracy in distance and speed detection. Besides, 77 GHz has a smaller antenna size, and it has less interference than 24 GHz. For 24 GHz radar, the maximum detection range is 70 m, while the maximum range increases to 200 m for 77 GHz radar. According to [93], the price for Continental's long-range radar can be around $3000, which is higher than the camera's price. However, compared with a camera, radar is less affected by the weather and low lighting environment, making it very useful in some applications like object detection and distance estimation. The data size is also smaller than the camera. Each radar produces 10–100 KB per second.

LiDAR Similar to Radar, LiDAR's distance information is also calculated based on the TOF. The difference is that LiDAR uses the laser for scanning, while radar uses electromagnetic waves. LiDAR consists of a laser generator and a high accuracy laser receiver. LiDAR generates a three-dimensional image of objects, so it is widely used to detect static objects and moving objects. LiDAR shows good performance with a range from several centimeters to 200 m, and the accuracy of distance goes to centimeter-level. LiDAR is widely used in object detection, distance estimation, edge detection, SLAM [519, 606], and High-Definition (HD) Map generation [135, 272, 293, 615]. Compared with the camera, LiDAR shows larger sensing range and its performance is less affected by bad weather and low lighting. However, in terms of the cost, LiDAR seems less competitive than camera and radar. According to [536], the 16 lines Velodyne LiDAR costs almost $8000, while the Velodyne VLS-128E costs over $100,000. High costs restrict the wide deployment of LiDAR on autonomous vehicles, contributing to the autonomous vehicle's high cost. LiDAR can generate almost 10-70MB data per second, a huge amount of data for the computing platform to process in real-time.

Ultrasonic Sensor Ultrasonic sensor is based on ultrasound to detect the distance. Ultrasound is a particular sound that has a frequency higher than 20 kHz. The distance is also detected by measuring TOF. The ultrasonic sensor's data size is close to the radar's, which is 10–100 KB per second. Besides, the ultrasonic sensor shows good performance in bad weather and low lighting environment. The ultrasonic sensor is much cheaper than the camera and radar. The price of the ultrasonic sensor is always less than $100. The shortcoming of ultrasonic sensors is the maximum range of only 20 m, limiting its application to short-range detection like parking assistance.

GPS/GNSS/IMU Except for sensing and perception of the surrounding environment, localization is also a significant task running on top of the autonomous driving system. In the localization system of the autonomous vehicle, GPS, GNSS, and IMU are widely deployed. GNSS is the name for all the satellite navigation systems, including GPS developed by the USA, Galileo from Europe, and BeiDou Navigation Satellite System (BDS) [40] from China. The accuracy of GPS can vary from several centimeters to several meters when different observation values and different processing algorithms are applied [180]. The strengths of GPS are low costs, and the non-accumulation of error over time. The drawback of GPS is that the GPS deployed on current vehicles only has accuracy within one meter: and GPS requires an unobstructed view in the sky, so it does not work in environments like tunnels, for example. Besides, the GPS sensing data updates every 100 ms, which is not enough for the vehicle's real-time localization.

IMU stands for inertial measurement unit, which consists of gyroscopes and accelerometers. Gyroscopes are used to measure the axes' angular speed to calculate the carrier's position. In comparison, the accelerometer measures the object's three axes' linear acceleration and can be used to calculate the carrier's speed and position. The strength of IMU is that it does not require an unobstructed view from the sky. The drawback is that the accuracy is low, and the error is accumulated with time. IMU can be a complementary sensor to the GPS because it has an updated value every 5 ms, and it works appropriately in environments like tunnels. Usually, a Kalman filter is applied to combine the sensing data from GPS and IMU to get fast and accurate localization results [316].

Table 1.2 shows a comparison of sensors, including camera, radar, LiDAR, and ultrasonic sensors with human beings. From the comparison, we can easily conclude that although humans have strength in the sensing range and show more advantaged application scenarios than any sensor, the combination of all the sensors can do a better job than human beings, especially in bad weather and low lighting conditions.

1.3.2 Data Source

Data Characteristics As we listed before, various sensors, such as GPS, IMU, camera, LiDAR, radar, are equipped in AVs, and they will generate hundreds of megabytes of data per second, fed to different autonomous driving algorithms. The data in AVs could be classified into two categories: real-time data and historical

Table 1.2 Comparisons of camera, radar, LiDAR, and ultrasonic sensor

Metrics	Human	Camera	Radar	LiDAR	Ultrasonic
Techniques	–	Lights	Electromagnetic	Laser reflection	Ultrasound
Sensing range	0–200 m	0–100 m	1 cm–200 m (77 GHz) 1 cm–70 m (24 GHz)	0.7–200 m	0–20 m
Cost	–	~$500	~$3000	$5000–$100,000	~$100
Data per second	–	20–40 MB	10–100 KB	10–70 MB	10–100 KB
Bad weather functionality	Fair	Poor	Good	Fair	Good
Low lighting functionality	Poor	Fair	Good	Good	Good
Application scenarios	Object detection Object classification Edge detection Lane tracking	Object classification Edge detection Lane tracking	Object detection Distance estimation	Object detection Distance estimation Edge detection	Object detection Distance estimation

data. Typically, the former is transmitted by a messaging system with the Pub/Sub pattern in most AVs solutions, enabling different applications to access one data simultaneously. Historical data includes application data. The data persisted from real-time data, where structured data, i.e., GPS, is stored into a database, and unstructured data, i.e., video, is stored as files.

Dataset and Benchmark Autonomous driving dataset is collected by survey fleet vehicles driving on the road, which provides the training data for research in machine learning, computer vision, and vehicle control. Several popular datasets provide benchmarks, which are rather useful in autonomous driving systems and algorithms design. Here are a few popular datasets: (1) *KITTI*: As one of the most famous autonomous driving datasets, the KITTI [162] dataset covers stereo, optical flow, visual odometry, 3D object detection, and 3D tracking. It provides several benchmarks, such as stereo, flow, scene, optical flow, depth, odometry, object tracking [217], road, and semantics [157]. (2) *Cityscapes*: For the semantic understanding of urban street scenes, the Cityscapes [507] dataset includes 2D semantic segmentation on pixel-level, instance-level, and panoptic semantic labeling and provides corresponding benchmarks on them. (3) *BDD100K*: As a large-scale and diverse driving video database, BDD100K [586] consists of 100,000 videos and covers different weather conditions and times of the day. (4) *DDD17*: As the first end-to-end dynamic and active-pixel vision sensors (DAVIS) driving dataset, DDD17 [48] has more than 12 h of DAVIS sensor data under different scenarios and different weather conditions, as well as vehicle control information like steering, throttle, and brake.

Labeling Data labeling is an essential step in a supervised machine learning task, and the quality of the training data determines the quality of the model. Here are a few different types of annotation methods: (1) *Bounding boxes*: the most commonly used annotation method (rectangular boxes) in object detection tasks to define the location of the target object, which can be determined by the x and y-axis coordinates in the upper-left corner and the lower-right corner of the rectangle. (2) *Polygonal segmentation*: since objects are not always rectangular, polygonal segmentation is another annotation approach where complex polygons are used to define the object's shape and location in a considerably precise way. (3) *Semantic segmentation*: a pixel-wise annotation, where every pixel in an image is assigned to a class. It is primarily used in cases where environmental context is essential. (4) *3D cuboids*: They provide 3D representations of the objects, allowing models to distinguish features like volume and position in a 3D space. (5) *Key-Point and Landmark* are used to detect small objects and shape variations by creating dots across the image. As to the annotation software, MakeSense.AI [341], LabelImg [280], VGG image annotator [538], LabelMe [281], Scalable [467], and RectLabel [432] are popular image annotation tools.

1.3.3   Autonomous Driving Applications

Plenty of algorithms are deployed in the computing system for sensing, perception, localization, prediction, and control. In this part, we present the state-of-the-art works for algorithms including object detection, lane detection, localization and mapping, prediction and planning, and vehicle control.

1.3.3.1   Object Detection

Accurate object detection under challenging scenarios is essential for real-world deep learning applications for AVs [355]. In general, it is widely accepted that the development of object detection algorithms has gone through two typical phases: (1) conventional object detection phase and (2) deep learning supported object detection phase [617]. Viola Jones Detectors [539], Histogram of Oriented Gradients (HOG) feature descriptor [96], and Deformable Part-based Model (DPM) [137] are all the typical traditional object detection algorithms. Although today's most advanced approaches have far exceeded the accuracy of traditional methods, many dominant algorithms are still deeply affected by their valuable insights, such as hybrid models, bounding box regression, etc. As to the deep learning-based object detection approaches, the state-of-the-art methods include the Regions with CNN features (RCNN) series [171, 172, 200, 441], Single Shot MultiBox Detector (SSD) series [152, 320], and You Only Look Once (YOLO) series [434–436]. Girshick et al. first introduced deep learning into the object detection field by proposing RCNN in 2014 [172, 173]. Later on, Fast RCNN [171] and Faster RCNN [441] were developed to accelerate detection speed. In 2015, the first one-stage object detector, i.e., YOLO was proposed [434]. Since then, the YOLO series algorithms have been continuously proposed and improved, for example, YOLOv3 [436] is one of the most popular approaches, and YOLOv4 [51] is the latest version of the YOLO series. To solve the trade-off problem between speed and accuracy, Liu et al. proposed SSD [320] in 2015, which introduces the regression technologies for object detection. Then, RetinaNet was proposed in 2017 [306] to further improve detection accuracy by introducing a new loss function to reshape the standard cross-entropy loss.

1.3.3.2   Lane Detection

Performing accurate lane detection in real-time is a crucial function of advanced driver-assistance systems (ADAS) [380], since it enables AVs to drive themselves within the road lanes correctly to avoid collisions, and it supports the subsequent trajectory planning decision and lane departure.

Traditional lane detection approaches (e.g., [55, 107, 228, 241, 508, 565]) aim to detect lane segments based on diverse handcrafted cues, such as color-based features [85], the structure tensor [323], the bar filter [511], and ridge features [324]. This information is usually combined with a Hough transform [308, 611] and particle or Kalman filters [97, 264, 511] to detect lane markings. Then, post-processing methods are leveraged to filter out misdetections and classify lane points to output the final lane detection results [208]. However, in general, they are prone to effectiveness issues due to road scene variations, e.g., changing from city scene to highway scene and hard to achieve reasonable accuracy under challenging scenarios without a visual clue.

Recently, deep learning-based segmentation approaches have dominated the lane detection field with more accurate performance [178]. For instance, VPGNet [287] proposes a multi-task network for lane marking detection. To better utilize more visual information of lane markings, SCNN [401] applies a novel convolution operation that aggregates diverse dimension information via processing sliced features and then adds them together. In order to accelerate the detection speed, light-weight DNNs have been proposed for real-time applications. For example, self-attention distillation (SAD) [219] adopts an attention distillation mechanism. Besides, other methods such as sequential prediction and clustering are also introduced. In [299], a long short-term memory (LSTM) network is presented to face the lane's long line structure issue. Similarly, Fast-Draw [414] predicts the lane's direction at the pixel-wise level. In [223], the problem of lane detection is defined as a binary clustering problem. The method proposed in [218] also uses a clustering approach for lane detection. Subsequently, a 3D form of lane detection [159] is introduced to face the non-flatten ground issue.

1.3.3.3 Localization and Mapping

Localization and mapping are fundamental to autonomous driving. Localization is responsible for finding ego-position relative to a map [278]. The mapping constructs multi-layer high-definition (HD) maps [238] for path planning. Therefore, the accuracy of localization and mapping affects the feasibility and safety of path planning. Currently, GPS-IMU based localization methods have been widely utilized in navigation software like Google Maps. However, the accuracy required for urban automated driving cannot be fulfilled by GPS-IMU systems [532].

Currently, systems that use a pre-build HD map are more practical and accurate. There are three main types of HD maps: landmark-based, point cloud-based, and vision-based. Landmarks such as poles, curbs, signs, and road markers can be detected with LiDAR [199] or camera [499]. Landmark searching consumes less computation than the point cloud-based approach but fails in scenarios where landmarks are insufficient. The point cloud contains detailed information about the environment with thousands of points from LiDAR [616] or camera [498]. Iterative closest point (ICP) [44] and normal distributions transform (NDT) [46] are two algorithms used in point cloud-based HD map generation. They utilize numerical optimization algorithms to calculate the best match. ICP iteratively selects the closest point to calculate the best match. On the other side, NDT represents the map as a combination of the normal distribution, then uses the maximum likelihood estimation equation to search match. NDT's computation complexity is less than ICP [342], but it is not as robust as ICP. Vision-based HD maps are another direction recently becoming more and more popular. The computational overhead limits its application in real systems. Several methods for matching maps with the 2D camera as well as matching 2D image to the 3D image are proposed for mapping [348, 413, 563].

In contrast, SLAM [59] is proposed to build the map and localize the vehicle simultaneously. SLAM can be divided into LiDAR-based SLAM and camera-based SLAM. Among LiDAR-based SLAM algorithms, LOAM [598] can be finished in real-time. IMLS-SLAM [105] focuses on reducing accumulated drift by utilizing a scan-to-model matching method. Cartographer [205], a SLAM package from Google, improves performance by using sub-map and loop closure while supporting both 2D and 3D LiDAR. Compared with LiDAR-based SLAM, camera-based SLAM approaches use frame-to-frame matching. There are two types of matching methods: feature-based and direct matching. Feature-based methods [370, 470, 500] extract features and track them to calculate the motion of the camera. Since features are sparse in the image, feature-based methods are also called sparse visual SLAM. Direct matching [128, 279, 382] is called dense visual SLAM, which adopts original information for matching that is dense in the image, such as color and depth from an RGB-D camera. The inherent properties of feature-based methods lead to its faster speed but tend to fail in texture-less environments as well. The dense SLAM solves the issues of the sparse SLAM with higher computation complexity. For situations that lack computation resources, semiDense [129, 430] SLAM methods that only use direct methods are proposed. Besides the above methods, deep learning methods are also utilized in solving feature extraction [576], motion estimation [302], and long-term localization [160].

1.3.3.4 Prediction and Planning

The prediction module evaluates the driving behaviors of the surrounding vehicles and pedestrians for risk assessment [589]. Hidden Markov model (HMM) has been used to predict the target vehicle's future behavior and detect unsafe lane change events [164, 573].

Planning means finding feasible routes on the map from origin to destination. GPS navigation systems are known as global planners [37] to plan a feasible global route, but it does not guarantee safety. In this context, the local planner is developed [176], which can be divided into three groups: (1) Graph-based planners that give the best path to the destination. (2) Sampling-based planners which randomly scan the environments and only find a feasible path. (3) Interpolating curve planners that are proposed to smooth the path. A* [197] is a heuristic implementation of Dijkstra that always preferentially searches the path from the origin to the destination (without considering the vehicle's motion control), which causes the planning generated by A* to not always be executed by the vehicle. To remedy this problem, hybrid A* [362] generates a drivable curve between each node instead of a jerky line. Sampling-based planners [251] randomly select nodes for search in the graph, reducing the searching time. Among them, Rapidly-exploring Random Tree (RRT) [283] is the most commonly used method for automated vehicles. As an extension of RRT, RRT* [252, 452] tries to search the optimal paths satisfying real-time constraints. How to balance the sampling size and computation efficiency is a big challenge for sampling-based planners. Graph-based planners and sampling-based planners can achieve optimal or sub-optimal with jerky paths that can be smoothed with interpolating curve planners.

1.3.3.5 Vehicle Control

Vehicle control connects autonomous driving computing systems and the drive-by-wire system. It adjusts the steering angle and maintains the desired speed to follow the planning module's trajectories. Typically, vehicle control is accomplished by using two controllers: lateral controller and longitudinal controller. Controllers must handle rough and curvy roads, and quickly varying types, such as gravel, loose sand, and mud puddles [215], which are not considered by vehicle planners. The output commands are calculated from the vehicle state and the trajectory by control law. There are various control laws, such as fuzzy control [9, 125], PID control [36, 419], Stanley control [215], and Model predictive control (MPC) [86, 235, 587]. PID control creates outputs based on proportional, integral, and derivative teams of inputs. Fuzzy control accepts continuous values between 0 and 1, instead of either 1 or 0, as inputs continuously respond. Stanley control is utilized to follow the reference path by minimizing the heading angle and cross-track error using a nonlinear control law. MPC performs a finite horizon optimization to identify the control command. Since it can handle various constraints and use past and current errors to predict more accurate solutions, MPC has been used to solve hard control problems like following overtaking trajectories [113]. Controllers derive control laws depending on the vehicle model. Kinematic bicycle models and dynamic bicycle models are most commonly used. In [271], a comparison is present to determine which of these two models is more suitable for MPC in forecast error and computational overhead.

1.3.4   Computation Hardware

To support real-time data processing from various sensors, powerful computing hardware is essential to autonomous vehicles' safety. Currently, plenty of computing hardware with different designs show up on the automobile and computing market. In this section, we will show several representative designs based on Graphic Processor Unit (GPU), Digital Signal Processor (DSP), Field Programmable Gate Arrays (FPGA), and Application-Specific Integrated Circuit (ASIC). The comparisons of GPU, DSP, FPGA, and ASIC in terms of architecture, performance, power consumption, and cost are shown in Table 1.3.

Table 1.3 The comparison of different computing hardware for autonomous driving

Boards	Architecture	Performance	Power consumption	Cost[a]
NVIDIA DRIVE PX2	GPU	30 TOPS	60 W	$15,000
NVIDIA DRIVE AGX	GPU	320 TOPS	300 W	$30,000
Texas Instruments TDA3x	DSP	–	30 mW in 30 fps	$549
Zynq UltraScale+ MPSoC ZCU104	FPGA	14 images/sec/Watt	–	$1295
Mobileye EyeQ5	ASIC	24 TOPS	10 W	$750
Google TPU v3	ASIC	420 TFLOPS	40 W	$8 per hour

[a]The cost of each unit is based on the price listed on their web site when the product is released to the market

NVIDIA DRIVE AGX is the newest solution from NVIDIA unveiled at CES 2018 [350]. NVIDIA DRIVE AGX is the world's most powerful System-on-Chip (SoC), and it is ten times more powerful than the NVIDIA Drive PX2 platform. Each DRIVE AGX consists of two Xavier cores. Each Xavier has a custom 8-core CPU and a 512-core Volta GPU. DRIVE AGX is capable of 320 trillion operations per second (TOPS) of processing performance.

Zynq UltraScale+ MPSoC ZCU104 is an automotive-grade product from Xilinx [127]. It is an FPGA-based device designed for autonomous driving. It includes 64-bit quad-core ARM® Cortex™-A53 and dual-core ARM Cortex-R5. This scalable solution claims to deliver the right performance/watt with safety and security [618]. When running CNN tasks, it achieves 14 images/s/W, which outperforms the Tesla K40 GPU (4 images/s/W). Also, for object tracking tasks, it reaches 60 fps in a live 1080p video stream.

Texas Instruments' TDA provides a DSP-based solution for autonomous driving. A TDA3x SoC consists of two C66x Floating Point VLIW DSP cores with vision AccelerationPac. Furthermore, each TDA3x SoC has dual Arm Cortex-M4 image processors. The vision accelerator is designed to accelerate the process functions on images. Compared with an ARM Cortex-15 CPU, TDA3x SoC provides an eight-fold acceleration on computer vision tasks with less power consumption [513].

MobileEye EyeQ5 is the leading ASIC-based solution to support fully autonomous (Level 5) vehicles [517]. EyeQ5 is designed based on 7 nm-FinFET semiconductor technology, and it provides 24Tops computation capability with 10 watts' power budget. TPU is Google's AI accelerator ASIC mainly for neural network and machine learning [11]. TPU v3 is the newest release, which provides 420 TFLOPS computation for a single board.

1.3.5 Storage

The data captured by an autonomous vehicle is proliferating, typically generating between 20 TB and 40 TB per day, per vehicle [143]. The data includes cameras (20–40 MB), as well as sonar (10–100 KB), radar (10–100 KB), and LiDAR (10–70 MB) [98, 514]. Storing data securely and efficiently can accelerate overall system performance. Take object detection, for example: the history data could contribute to the improvement of detection precision using machine learning algorithms. Map generation can also benefit from the stored data in updating traffic and road conditions appropriately. Additionally, the sensor data can be utilized to ensure public safety and predict and prevent crime. The biggest challenge is to ensure that sensors collect the right data, and it is processed immediately, stored securely, and transferred to other technologies in the chain, such as Road-Side Unit (RSU), cloud data center, and even third-party users [600]. More importantly, creating hierarchical storage and workflow that enables smooth data accessing and computing is still an open question for the future development of autonomous vehicles.

In [451], a computational storage system called HydraSpace is proposed to tackle the storage issue for autonomous driving vehicles. HydraSpace is designed with multi-layered storage architecture and practical compression algorithms to manage the sensor pipe data. OpenVDAP is a full-stack edge-based data analytic platform for connected and autonomous vehicles (CAVs) [600]. It envisions for the future four types of CAVs applications, including autonomous driving, in-vehicle infotainment, real-time diagnostics, and third-party applications like traffic information collector and SafeShareRide [313]. The hierarchical design of the storage system called driving data integrator (DDI) is proposed in OpenVDAP to provide sensor-aware and application-aware data storage and processing [600].

1.3.6   Real-Time Operating Systems

According to the automation level definitions from the SAE [478], the automation of vehicles increases from level 2 to level 5, and the level 5 requires full automation of the vehicle, which means the vehicle can drive under any environment without the help from the human. To make the vehicle run in a safe mode, how to precept the environment and make decisions in real-time becomes a big challenge. That is why real-time operating systems become a hot topic in the design and implementation of autonomous driving systems.

RTOS is widely used in the embedded system of ECUs to control the vehicle's throttle, brake, etc. *QNX* and *VxWorks* are two representative commercialized RTOS widely used in the automotive industry. The *QNX* kernel contains only CPU scheduling, inter-process communication, interrupt redirection, and timers. Everything else runs as a user process, including a unique process known as "proc," which performs process creation and memory management by operating in conjunction with the microkernel [206]. *VxWorks* is designed for embedded systems requiring real-time, deterministic performance and, in many cases, safety and security certification [541]. *VxWorks* supports multiple architectures, including Intel, POWER, and ARM. *VxWorks* also uses real-time kernels for mission-critical applications subject to real-time constraints, which guarantees a response within predefined time constraints.

RTLinux is a microkernel-based operating system that supports hard real-time [584]. The scheduler of *RTLinux* allows full preemption. Compared with using a low-preempt patch in *Linux*, *RTLinux* allows preemption for the whole Linux system. *RTLinux* makes it possible to run real-time critical tasks and interprets them together with the *Linux* [460].

NVIDIA DRIVE OS is a foundational software stack from NVIDIA, which consists of an embedded RTOS, hypervisor, NVIDIA CUDA libraries, NVIDIA Tensor RT, etc. that is needed for the acceleration of machine learning algorithms [390].

1.3.7   Middleware Systems

Robotic systems, such as autonomous vehicle systems, often involve multiple services, with many dependencies. Middleware is required to facilitate communications between different autonomous driving services.

Most existing autonomous driving solutions utilize the *ROS* [424]. Specifically, *ROS* is a communication middleware that facilitates communications between different modules of an autonomous vehicle system. *ROS* supports four communication methods: topic, service, action, and parameter. *ROS2* is a promising type of middleware developed to make communications more efficient, reliable, and secure [448]. However, most of the packages and tools for sensor data process are still currently based on *ROS*.

The *Autoware Foundation* is a non-profit organization supporting open-source projects enabling self-driving mobility [20]. *Autoware.AI* is developed based on *ROS*, and it is the world's first "all-in-one" open-source software for autonomous driving technology. *Apollo Cyber* [30] is another open-source middleware developed by Baidu. *Apollo* aims to accelerate the development, testing, and deployment of autonomous vehicles. *Apollo Cyber* is a high-performance runtime framework that is greatly optimized for high concurrency, low latency, and high throughput in autonomous driving.

In traditional automobile society, the runtime environment layer in Automotive Open System Architecture(*AutoSAR*) [19] can be seen as middleware. Many companies develop their middleware to support *AutoSAR*. However, there are few independent open-source middleware nowadays because it is a commercial vehicle company's core technology. Auto companies prefer to provide middleware as a component of a complete set of autonomous driving solutions.

1.3.8   Vehicular Communication

In addition to obtaining information from the on-board sensors, the recent proliferation in communication mechanisms, e.g., DSRC, C-V2X, and 5G, has enabled autonomous driving vehicles to obtain information from other vehicles, infrastructures like traffic lights and RSU as well as pedestrians.

LTE/4G/5G Long-Term Evolution (LTE) is a transitional product in the transition from 3G to 4G [312], which provides downlink peak rates of 300 Mbit/s, uplink peak rates of 75 Mbit/s. The fourth-generation communications (4G) comply with 1 Gbit/s for stationary reception and 100 Mbit/s for mobile. As the next-generation mobile communication, U.S. users that experienced the fastest average 5G download speed reached 494.7 Mbps on Verizon, 17.7 times faster than that of 4G. And from Verizon's early report, the latency of 5G is less than 30 ms, 23 ms faster than average 4G metrics. However, we cannot deny that 5G still has the following challenges: complex system, high costs, and poor obstacle avoidance capabilities.

DSRC DSRC [258] is a type of V2X communication protocol, which is specially designed for connected vehicles. DSRC is based on the IEEE 802.11p standard, and its working frequency is 5.9 GHz. Fifteen message types are defined in the SAE J2735 standard [492], which covers information like the vehicle's position, map information, emergence warning, etc. [258]. Limited by the available bandwidth, DSRC messages have small size and low frequency. However, DSRC provides reliable communication, even when the vehicle is driving 120 miles per hour.

C-V2X C-V2X combines the traditional V2X network with the cellular network, which delivers mature network assistance and commercial services of 4G/5G into autonomous driving. Like DSRC, the working frequency of C-V2X is also the primary common spectrum, 5.9 GHz [18]. Different from the CSMA-CA in DSRC, C-V2X has no contention overheads by using semi-persistent transmission with relative energy-based selection. Besides, the performance of C-V2X can be seamlessly improved with the upgrade of the cellular network. Generally, C-V2X is more suitable for V2X scenarios where cellular networks are widely deployed.

1.3.9 Security and Privacy

With the increasing degree of vehicle electronification and the reliance on a wide variety of technologies, such as sensing and machine learning, the security of AVs has risen from the hardware damage of traditional vehicles to comprehensive security with multi-domain knowledge. Here, we introduce several security problems strongly associated with AVs with the current attacking methods and standard coping methods. In addition to the security and privacy issues mentioned as follows, AVs systems should also take care of many other security issues in other domains, such as patching vulnerabilities of hardware or software systems and detecting intrusions [551].

Sensing Security As the eye of autonomous vehicles, the security of sensors is nearly essential. Typically, jamming attacks and spoofing attacks are two primary attacks for various sensors [440, 580]. For example, the spoofing attack generates an interference signal, resulting in a fake obstacle captured by the vehicle [574]. Besides, GPS also encounters spoofed attacks [592]. Therefore, protection mechanisms are expected for sensor security. Randomized signals and redundant sensors are usually used by these signal-reflection sensors [412, 485], including LiDAR and radar. The GPS can check signal characteristics [273] and authenticate data sources [392] to prevent attacks. Also, sensing data fusion is an effective mechanism.

Communication Security Communication security includes two aspects: internal communication and outside communication. Currently, internal communication, like CAN, LIN, and FlexRay, has faced severe security threats [131, 275, 386]. The cryptography is frequently used technology to keep the transmitted data confidential, integrated, and authenticated [495]. However, the usage of cryptography is limited by the high computational cost for these resource-constrained ECUs. Therefore, another attempt is to use the gateway to prevent unallowed access [263]. The outside communication has been studied in VANETs with V2V, V2R, and V2X communications [7, 374, 423]. Cryptography is the primary tool. A trusted key distribution and management is built in most approaches, and vehicles use assigned keys to authenticate vehicles and data.

Data Security Data security refers to preventing data leakage from the perspectives of transmission and storage. The former has been discussed in communication security, where various cryptography approaches are proposed to protect data in different scenarios [158, 609]. The cryptography is also a significant technology of securing data storage, such as an encrypted database [418] and file system [50]. Besides, access control technology [457] protects stored data from another view, widely used in modern operating systems. An access control framework [604] has been proposed for AVs to protect in-vehicle data in real-time data and historical data, with different access control models.

Control Security With vehicles' electronification, users could open the door through an electronic key and control their vehicles through an application or voice. However, this also leads to new attack surfaces with various attack methods, such as jamming attacks, replay attacks, relay attacks, etc. [440]. For example, attackers could capture the communication between key and door and replay it to open the door [242]. Also, for those voice control supported vehicles, the attackers could successfully control the vehicle by using voices that humans cannot hear [596]. Parts of these attacks could be classified into sensing security, communication security, or data security, which can be addressed by corresponding protection mechanisms.

Privacy Autonomous vehicles heavily rely on the data of the surrounding environment, which typically contains user privacy. For example, by recognizing buildings in cameras, attackers can learn the vehicle location [570]. Or an attacker can obtain the location directly from GPS data. Thus, the most straightforward but the most difficult solution is to prevent data from being obtained by an attacker, such as access control [457, 604] and data encryption [440]. However, autonomous vehicles will inevitably utilize location-based services. Except for the leak of current location, attackers could learn the home address from the vehicle trajectory [297]. Thus, data desensitization is necessary to protect privacy, including anonymization and differential privacy [345].

1.4 Overview of the Book

The remaining chapters covered in this book include the following:

- **Computing Framework for Autonomous Driving.** Software stack plays a significant role in autonomous vehicle's computing system. To build a robust software stack, a sophisticated understanding of the application requirements is necessary. Chapter 2 presents typical computing frameworks that includes OpenVDAP, HydraSpace, and AC4AV.

- **Algorithm Deployment Optimization.** With the burgeoning growth of the Internet of Everything(IoE), the amount of data generated by these edge devices has increased dramatically. Terabytes of sensor data could be generated per vehicle per day, which makes the algorithm deployment become a big challenge. Chapter 3 presents CLONE and driving behavior modeling to show the algorithm deployment optimizations.
- **Systems Runtime Optimization.** As a safety-critical system, how to satisfy the real-time requirements of CAVs applications becomes an essential challenge. Besides, since the on-board battery powers the whole computing system, the computing system's power dissipation can be a big issue. These system requirements make runtime optimization become significant. Chapter 4 presents E2M and determinism analysis of DNN inference to show the runtime optimizations.
- **Dataset and Benchmark.** Autonomous driving dataset is collected by survey fleet vehicles driving on the road, which provides the training data for research in machine learning, computer vision, and vehicle control. Chapter 5 presents open dataset and benchmark for autonomous driving.
- **Autonomous Driving Simulator.** The deployment of autonomous driving algorithms or prototypes requires complex tests and evaluations in a real environment, which makes it necessary to build powerful and realistic simulators. Chapter 6 present the autonomous driving simulators.
- **Hardware Platforms.** Although simulators provide a rich scenario for testing and validation, the real road environment is more complicated than simulators. For the safety of autonomous driving vehicles, building hardware platforms for the vehicles as well as the traffic infrastructure become essential. Chapter 7 presents several typical autonomous driving experimental platform, including HydraOne, Hydra, and Equinox.
- **Smart Infrastructure for Autonomous Driving.** In addition to the computing system design, Vehicle-to-everything (V2X) technologies should also get involved for the communication of edge system. Chapter 8 presents the current status of V2X communications.
- **Challenges and Open Problems.** Although plenty of innovations develop in autonomous driving technologies, there are still many challenges and open issues for the research and development of L4 or L5 autonomous driving vehicles. Chapter 9 presents the remaining challenges for autonomous driving.

Chapter 2
Computing Framework for Autonomous Driving

2.1 In-Vehicle Applications

In this section, we briefly discuss four types of applications that will be available on CAVs. Conventionally, these services on current vehicles could be classified into three groups according to their functionality: *real-time diagnostics*, *advanced driver-assistant systems*, and *in-vehicle infotainment*. In addition, we envision a new type of services, third-party applications from various vendors, will be prevalent on CAVs as the vehicle data in the future will not be exclusive to the auto-makers.

2.1.1 Real-Time Diagnostics

This type of service usually refers to the On-board diagnostics (OBD) system, which allows the vehicle to have the capability of self-diagnosis and reporting. The OBD system appeared on the vehicle in the 1980s and has evolved from the early simple "idiot light" to a modern version that can provide real-time vehicle data (e.g., the engine's revolution from the engine control unit) and a standardized series of diagnostic trouble codes. Such code is useful for vehicle maintenance and repair. The device reading the real-time data is actually an additional device to the vehicle called the OBD reader. The maintainer can leverage the OBD reader to obtain information about the fault, e.g., the diagnostic trouble code. Thus, this usually will not consume any resources of the vehicle. However, it is not an in-vehicle system. In future CAVs, this type of service should be built in the vehicle, which collects the related vehicle data, including real-time data and historical data, and quietly analyzes it to predict faults. Thus, it can remind the owner of keeping the vehicle in good condition.

W. Shi, L. Liu, *Computing Systems for Autonomous Driving*,
https://doi.org/10.1007/978-3-030-81564-6_2

2.1.2   Advanced Driver-Assistant Systems

Nowadays, more and more vehicles are equipped with the ADAS that can detect some objects, complete basic classification, and alert the driver of unsafe driving behaviors. It may also slow or stop the vehicle. For example, the steering wheel will vibrate when the vehicle is close to a traffic line without illuminating the turn light. The Society of Automotive Engineers (SAE) International grades autonomous driving at six levels [233], in which the level 0 means the vehicle is without any assistant, and the level 5 means the vehicle is in full control by an autonomous driving system. A vehicle with ADAS is usually rated the SAE level 1 or level 2 based on provided functions. Usually, level 3 means the human driver can safely turn their attention away from driving tasks. As the level ascends, the autonomous driving system takes over more controls from human drivers and needs more computation resources to run computationally intensive algorithms.

The most important element of ADAS is the real-time object detection that is based on either computer vision or deep learning technology. To obtain intuitional perception on detection algorithm's computation complexity, we chose two of the most representative detection algorithms: Lane Detection and Vehicle Detection. The former relies on the computer vision technology. Regarding the latter, the underlying technologies are Haar-based image processing and TensorFlow-based deep learning algorithm. We conducted our experiments on the AWS EC2node with 2.4 GHz vCPU. For the vehicle detection, our initial results, as shown in Table 2.1, show that the latency of Haar-based algorithm significantly outperforms (around 51x faster) than the TensorFlow-based. But in the future CAVs, deep learning-based algorithms will dominate since they can detect multiple types of objects at once. To reduce the system latency, more powerful hardware is highly required.

2.1.3   In-Vehicle Infotainment

In-Vehicle infotainment includes a wide range of services that provide audio or video entertainment. For example, the driver uses the radio to listen to music, including cloud services from the Internet such as Pandora, and the passengers sitting in the backseat can relax by watching online videos or news using the device embedded on the seat. This means these services might involve large-scale Internet data transmission. Most mainstream auto vendors have Internet-supported infotainment services, e.g., Uconnect for Chrysler, Blue Link for Honda, and iDrive

Table 2.1 The performance of autonomous driving-related algorithms

Name	Latency (ms)
Lane detection	13.57
Vehicle detection (Haar)	269.46
Vehicle detection (TensorFlow)	13971.98

for BMW. Another trend of on-board infotainment is the Android-based system that has been implemented by many auto vendors including Honda, Hyundai, Audi, and Volvo. For these services, video or audio data must be downloaded from the Internet and then decoded locally (i.e., on the vehicle) to make the data smooth or at higher quality, and eventually delivered to the passengers.It means these applications not only require compute resources but also present a high requirement on the network bandwidth.

2.1.4   Third-Party Applications

The in-vehicle third-party application provided by a vendor other than the car maker is used to enhance the user experiences or provide other add-on services (e.g., finding a missing car for law enforcement). The trend of openness of vehicle data will make it easier for third-party vendors to develop and deploy different kinds of applications on the vehicle. In addition, with the rapid development of in-vehicle processor processing, vehicle-to-vehicle communications, and autonomous driving technologies, future CAVs could be viewed as a sophisticated computer on wheels with a variety of third-party applications.

Several projects including these types of applications have been initiated. For example, Kar et al. [248] proposed an application to enhance the vehicle safety by detecting whether the driver is registered or not through analyzing his/her operation features (e.g., the time duration of door open and close). Another example is to leverage the on-board camera to recognize and track a targeted vehicle, which is a mobile version for A3 [601] promising to enhance the AMBER alert system. Generally, the core of such types of applications is either vision-related or machine learning-based data processing algorithm fed by the same on-board sensor data (e.g., dash camera). Since the core algorithms are computationally intensive, this type of third-party applications needs powerful computing hardware as well.

The widespread of CAVs will also be a key component of smart homes to assist people's daily life. For example, CAVs can provide a new, convenient, and secure in-vehicle delivery service when the customer is away from home. Today, Amazon, the world's largest online retailer, is taking the obvious next step by cooperating with mainstream auto-makers and launching early in-vehicle delivery services. Once the delivery driver reaches the vehicle parked in a publicly accessible place, the driver will send a request to remotely unlock the vehicle for delivery. After placing packages in cargo area or cabin, the driver will send a remote command to lock the vehicle again, and the customer will receive a final notification. Following this way, customers can receive packages safely even when they are not home.

In addition, since fully autonomous driving could achieve safe and reliable navigation by itself, there is no driver needed to focus on driving anymore. In this context, future CAVs are expected to provide other intelligent services such as providing efficient and smooth online meeting experiences in vehicles. Specifically, as the rapid development of wireless and sensor technologies enables secure and interoperable communications among vehicles, clouds, and devices/things (such as passengers' personal communication devices), we envision that the future CAVs are able to support in-vehicle meetings allowing people to share information and data without being physically present at home or office. Besides, people will be able to seamlessly attend the same meeting at home, in the vehicle, and in the office without being bothered by the repeated logout and login process, which really improves work efficiency and saves working time.

2.2 OpenVDAP: An Open Vehicular Data Analytics Platform for CAVs

2.2.1 Introduction

As the rapid growth of technologies in communication, chip processing, sensing and machine learning, vehicles are becoming increasingly connected and automated [15, 319, 515, 554]. A typically connected and autonomous vehicle is often equipped with a High-Definition (HD) map that provides CAVs with detailed road data, such as the road shoulders, and a plethora of diverse sensors, e.g., light detection and ranging (LiDAR), radio detection and ranging (radar) and camera, to monitor the vehicle itself and its surroundings. According to Intel, a CAV will generate 4TB of data per day in the near future [378]. Although the sensor data can be transmitted to and processed in a remote cloud, the bandwidth and latency challenges make it impossible to ensure the real-time service (i.e., autonomous driving) provides timely decision all the time. Thus, the computing solution adopted by the autonomous driving platform is to process data on the vehicles themselves.

Keeping data processing on the vehicle is effective in offering real-time services. However, doing this will impose a huge challenge on the on-board computing unit, especially for the future CAVs that include more sensors and third-party applications with varying degree of computation demands. One naive solution is to add extra computing chips (e.g., GPU or DSP) once the on-board compute resource is not sufficient to support upper services. Since the powerful processors are often power-hungry, usually on the order of hundreds of watts for GPU, this method will cause processor overheating and a concern for energy consumption.

In this chapter, we envision the future CAVs as a sophisticated computer on wheels with a variety of services, such as advanced driver-assistant systems (ADAS), remote real-time diagnostics, in-vehicle infotainment, other third-party applications running on top. These applications could be the embedded services on the vehicle or newly added third-party services. Note that they are usually based on large-scale data analytic, thus quite computation expensive. For the in-vehicle only computing model, these applications need to contend with each other for the limited on-board computing resources. For instance, assume two latency-sensitive applications require execution on the GPU at the same time. If there is only one

Fig. 2.1 Data processing
model in an edge-based
solution

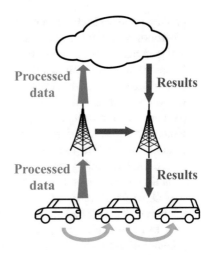

GPU on the vehicle, the second scheduled application might not produce a *timely* decision.

The emerging edge computing [461, 483] (also referred to as fog computing [54], cloudlet [462, 463]) is a novel networked computing architecture which will deploy the compute, storage, and communication resource at the network edge, enabling the latency-sensitive and confidentiality-aware applications to be performed in proximity of the data source, thus removing extra data transmission time and potential security and privacy risk. As shown in Fig. 2.1, the vehicles are able to communicate with each other and send processed data to nearby (usually one-hop away) edge servers (i.e., base stations) located between remote cloud and edge vehicles, with more powerful compute resources than the on-board computing unit, at the same time, producing much smaller latency and less chance in privacy leaking, compared with the cloud solution.

Research on CAV technologies is one of the hottest areas. In addition, machine learning-based real-time applications dominate on CAVs. Unfortunately, there are no edge computing platforms that support vehicular data analytic and processing. Many companies, such as Ford [146], General Motors [351], and Baidu, are working on this. Except for Baidu's Apollo [15], they are all proprietary. Though Apollo is open-source, it relies on the in-vehicle only computing solution that puts all the data processing on the vehicle, which is neither scalable nor suitable for the future CAVs with plenty of third-party services.

Motivated by the above observations, we propose to build an edge computing based Open Vehicular Data Analytics Platform (*OpenVDAP*) for future CAVs which is a full-stack edge supported platform that includes a series of heterogeneous hardware and software solutions. The key characteristics of *OpenVDAP* are summarized below:

- *OpenVDAP* supports on-board computing environment by providing multiple carefully selected heterogeneous computing processors and a security and privacy-preserved vehicle operating system to ensure a safe and trusted execution environment for upper applications all while maintaining effective and optimal resource management and utilization for lower diverse hardware.

- In addition to the on-board computing, *OpenVDAP* is two-tier computing architecture based. In particular, *OpenVDAP* provides systematic mechanisms on how to request, utilize, share, and even collaborate with external computing entities located on neighboring vehicles, nearby roadside edge servers, or remote cloud servers. And a dynamic offloading and scheduling algorithm is included to allow *OpenVDAP* to detect each service's status, computation overhead, and the optimal offloading destination so that each service could be completed at the right time with limited bandwidth consumption.

- The *OpenVDAP* expected to run on an autonomous vehicle test-bed offers an open and free edge-aware application library named libvdap that contains how to access and deploy edge computing based vehicle applications, various commonly used artificial intelligent (AI) models, as well as the interface of accessing open real field vehicle data, all of which will enable the researchers, developers in the community to freely deploy, test, and validate their applications in the real environment.

2.2.2   Overview of OpenVDAP

In this section, we will introduce Open Vehicular Data Analytics Platform (*Open-VDAP*), an edge-based solution for future CAVs. Considering the constraints of network (i.e., latency, bandwidth, and connectivity) and on-board computing resources, *OpenVDAP* enables the vehicle to collaborate with surrounding vehicles, offloading workloads to edge servers, denoted as *XEdge*, which could be running on base stations, RSUs, and traffic signal systems, as well as on remote Cloud. Below is an overview of *OpenVDAP* followed by a detailed discussion of each component.

As is shown in Fig. 2.2, *OpenVDAP* is a full-stack vehicle computing platform for future CAVs with the capabilities of collaborating with other edge nodes (i.e., nearby vehicles), XEdge, and a remote Cloud. It consists of four in-vehicle components: an on-board heterogeneous computing/communication unit (VCU), an isolation-supported and security/privacy-aware vehicle operating system (EdgeOS$_v$), a driving data integrator (DDI), and an edge-aware application library (libvdap),

Specifically, the VCU is a heterogeneous computing unit in terms of processors, storage, and communication modules. On top of VCU, it is an operating system, EdgeOS$_v$, which is responsible for providing a security running environment and energy-efficient and latency-aware resource management for upper applications. A library, libvdap, is provided allowing the third-party developers to build their own applications on the *OpenVDAP*. Additionally, *OpenVDAP* also contains a service called DDI that can collect all driving data generated by the in-vehicle sensors (e.g., OBD and dash camera) and other external information (e.g., local weather

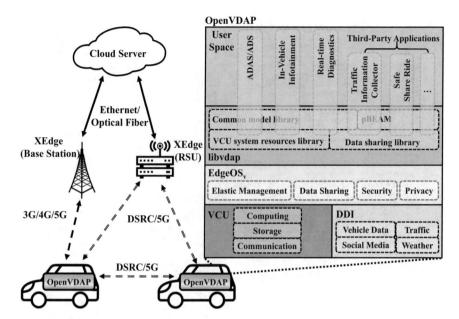

Fig. 2.2 The overview of *OpenVDAP*

from the Internet). All data collected by the DDI will be cached on the vehicle and eventually migrated to a cloud-based data server. Note that these data will be open to the community.

To enable collaboration between vehicles, XEdge and a remote Cloud, *Open-VDAP* supports multiple wireless interfaces such as DSRC, 5G, 3G/4G/LTE, WiFi, and Bluetooth, in which the DSRC and 5G will be used for vehicle-to-vehicle and vehicle to RSU-based XEdge communication due to their higher bandwidth. In addition, the vehicle can communicate with the base station via traditional cellular networks (e.g., 3G/4G/LTE). As the communication between the RSU/base station and cloud, the wired Ethernet or Optical Fiber will be used.

2.2.3 VCU: Heterogeneous Vehicle Computing Unit

Traditionally, vehicles usually contain a computing unit on board, also known as on-board controller. However, this on-board controller does not provide any open interface to other users/developers. In general, it has very limited computing power, failing to support the state-of-the-art applications, such as autonomous driving. Hence, we propose a new computing platform for vehicles called *Heterogeneous Vehicle Computing Unit (VCU)*, mainly consisting of a *multi-level heterogeneous computing platform (mHEP)* and a *dynamic scheduling framework (DSF)*. The architecture of VCU is shown in Fig. 2.3, in which mHEP relies on heterogeneous

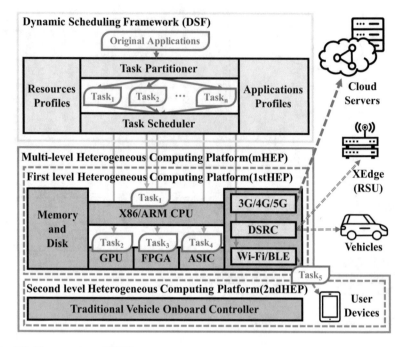

Fig. 2.3 The overview of VCU

hardware devices to provide the functions of computing, communication, and storage, and DSF to effectively utilize and manage the underlying hardware resources, especially the heterogeneous processors. The mHEP and DSF are described in detail below.

mHEP: Multi-Level Heterogeneous Computing Platform The *first level heterogeneous computing platform (1stHEP)* is the computing platform on the vehicle that contains heterogeneous hardware. This level is the main computing resource of VCU. In addition to CPU, 1stHEP leverages GPU, FPGA, and ASIC to match and accelerate the services on the vehicle. Due to the excellence in paralleled computation that benefits from thousands of hard floating point units, GPU is becoming widely used for accelerating complex machine learning or computer vision-based applications [90]. FPGA can be dynamically reconfigured, so it is suitable for various algorithms, including but not limited to machine learning [72, 156]. FPGA will perform the tasks like feature extraction, and data compression and media coding and decoding, etc. The 1stHEP will use some ASICs to accelerate specific algorithms, because they have best performance and energy efficiency.

The storage and communication modules are also deployed on the 1stHEP. In order to achieve better I/O performance, the parallelism-supported solid state drive (SSD) [4, 377] is chosen to store vehicle data and applications. Regarding the communication module, 1stHEP also contains several communication modules, such as 3G/4G/5G, Wi Fi/BLE, and DSRC. The 3G/4G/5G modules enable the vehicles to communicate with the cloud server directly. Vehicles are able to join the Internet of Vehicles through the 5G/DSRC module to cooperate or data share with other vehicles or XEdge. And the short-range Bluetooth interface will allow the vehicle to connect with the passengers' mobile devices.

Additionally, VCU also contains a *second level heterogeneous computing platform (2ndHEP)* that tries to exploit all other possible on-board computing resources, such as the passengers' mobile devices and vehicle on-board controller to alleviate the computation burden on the 1stHEP.

DSF: Dynamic Scheduling Framework The upper applications on the vehicle are often computational expensive. Meanwhile, they may depend on different kinds of computation tasks, such as data pre-possessing, model training/refining and decision prediction, each requiring different compute resources. We can assume the upper computation tasks are heterogeneous as well. Note that these tasks are not fixed over time, as do the underlying available computing resource. To optimally utilize the hardware resource, the primary goal of DSF is to provide an effective scheduling framework to dynamically assign each task/sub-task to the best *fit* processor. In particular, DSF has the following two functions:

Computing Resources Collection DSF provides dynamic management of computing resources of mHEP. First, DSF allows computing resources to join and exit dynamically, which is used to manage the 2ndHEP and some plug-and-play computing resources. Second, DSF acquires the real-time status of all computing resources periodically. For example, utilization rate and task type at processing. These dynamic status and static information (computing ability and matched task type) of computing resources are taken as their profiles. Third, the profiles are important information for DSF to dynamic allocate computing resources to applications. And resources accessed by applications are tightly controlled by DSF, which will achieve resources isolation and reduce the interference among the applications. DSF also provides the access interfaces of all computing resources, which we called *control knob*. The upper system or applications can access the computing resources via the control knob.

Task Scheduling DSF divides the original applications into some sub-tasks by fine-grained and tries to match the tasks with the computing resources according to their computing characteristics. DSF determines the resources type and amounts which will be allocated to each task according to the dynamic status of each resource, QoS requirement and processing priority of each task, and the cost of each scheduling plan. After the tasks are distributed to specified computing resources, DSF will reduce the results of each task and return it to the upper operating system or application.

Note that VCU could be viewed as complementary to the traditional vehicle on-board controller by offering more computing, storage, and communication capabilities for modern vehicle applications. The resources on VCU can be accessed by users, auto vendors, and third-party developers via the open interfaces. Moreover, the hardware resources could be easily augmented through the extensible interfaces (USB/PCIe).

The main design question for VCU is how to build an efficient computing platform for a vehicle that supports a wide array of applications on vehicle scenario. This is challenging for various reasons. First, VCU will integrate heterogeneous hardware (CPU, GPU, FPGA, and ASIC) on a single board. And, the board will contain several communication modules to make the vehicle connected to cloud, XEdge, other vehicles, and user devices. Second, VCU needs to provide dynamic management to resources and collect the real-time status of resources which will ensure the proper resource allocation for tasks. Finally, because all resource allocations and task distributions depend on the scheduling algorithm in VCU, the algorithm should consider more possible factors to make the best scheduling plan. Meanwhile, the complexity and overhead of the scheduling algorithm should be considered.

2.2.4 EdgeOSv: An Edge Operating System for Vehicles

EdgeOS$_v$ is an edge operating system for the CAVs, consisting of *Elastic Management*, *Data Sharing*, *Security*, and *Privacy*. Four fundamental features (DEIR) of service quality are proposed in [67, 483] for Edge Computing, which also should be inherited in the design of EdgeOS$_v$. The DEIR refers to *Differentiation, Extensibility, Isolation* and *Reliability*. *Differentiation* is achieved by the *Elastic Management* module and *Isolation* by the *Security* module. *Reliability* is supported together by two modules, *Elastic Management* and *Security*. It is worth noting that the *Extensibility* property is two-fold: hardware level and software level. The hardware level of extensibility is achieved by VCU, which can extend the computational devices using mobile devices, while the software level of extensibility is achieved by an application library called libvdap discussed in Sect. 2.2.6.

The *Elastic Management* module is used to manage all services on the vehicle. The *Data Sharing* module provides the data exchange and sharing between different services. And the *Security* module is used to provide a trusted execution environment and isolation scheme for security-sensitive services. The *Privacy* module provides some privacy protection schemes for data sharing between vehicles and XEdge, as well as the cloud. Before we introduce these four modules, we first introduce the conception of *Polymorphic service* and then introduce how the *Elastic Management* module manages these services. Then, we will introduce *Data Sharing, Security*, and *Privacy* modules.

In EdgeOS$_v$, each service offers multiple execution pipelines in response to various network and computational constraints. Take the third-party application searching for a kidnapper using mobile version A3, as an example. The vehicle can automatically pinpoint the targeted vehicle by recognizing the license plate number. This service can be accomplished in three pipelines: (1) all workloads execute on board; (2) all workloads execute on the edge/cloud server, and (3) the detecting

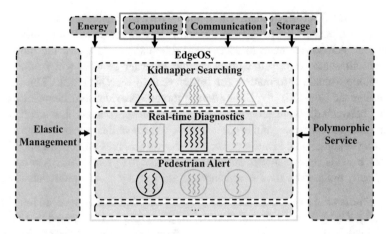

Fig. 2.4 The overview of *Elastic Management*

motion is put on board while recognizing license plate is deployed on the edge/cloud server. Here, the first one needs more local computational resources than the second, but with less bandwidth requirement and low network latency.

The *Elastic Management* module is illustrated in Fig. 2.4. The *Elastic Management* module can dynamically choose an optimal pipeline for each *Polymorphic Service*, considering priority, required response time, and polymorphic requirements for computational resource and network quality (i.e., latency, bandwidth, and connectivity). Let us take the kidnapper searching as an example, if the network quality (i.e., remainder bandwidth by other applications) is enough for uploading video data the nearby XEdge, the *Elastic Management* module might offload all workloads to the XEdge, especially when there is a lack of on-board computational resources. Once all of bandwidth are occupied by other services with higher priority (e.g., autonomous driving services), it can adjust the pipeline to make part of workload being processed on board, e.g., detecting motion. If the network quality and computation resources cannot support this service, the service will be hung up until meeting requirements again. As the dynamic adjustment, the service quality and user experience will be optimized.

To address the security issues of CAVs, EdgeOS$_v$ also contains a *Security* module, which relies on the trusted execution environment (TEE) technique. The major benefits of using TEE can ensure all services running on top be securely isolated via encryption of their corresponding memory space. For other non-TEE supported services, the containerization, compared with the virtualization technology, is a good candidate for isolation and migration due to the lightweight of a container [122]. Note that the containerization also can enhance the isolation of a TEE [17]. Moreover, the *Security* module monitors services and prevents them from compromising. Once the service is compromised , this module will remove the compromised one and re-install an initialized one without compromising, which implements the part of function of *Reliability*.

As mentioned before, the data will be shared between vehicles as well as services. For example, both of the pedestrian detection service for autonomous driving and the mobile A3 service need to access real-time camera data, and the mobile A3 service will share the result with a vehicle recorder service, which records all surrounding vehicle information for future vehicle searching task. Thus, in our TEE enhanced EdgeOS$_v$, the *Data Sharing* module provides a mechanism for data sharing between different services with a high security, which will authenticate the service and perform fine grain access control. To protect the privacy of data sharing between vehicles, some identity privacy protection schemes will be provided by the *Privacy* module. For example, the vehicle can use the pseudonym, generated and periodically updated by the *Privacy* module, for privacy protection in data sharing.

Open Problems Here are several open problems that need to be addressed for EdgeOS$_v$. Zhang et al. [602] and Kang et al. [244] have demonstrated that dividing a workload into several parts and making them execute on different edge nodes along the path from the source to the cloud can get a better response latency and data transmission. However, how to dynamical divide workload on the edges is still a problem. And in our EdgeOS$_v$, it requires knowing the network quality to other edge nodes, which has not been well solved.

2.2.5 DDI: Driving Data Integrator

Besides the data collected by sensors embedded in different components of vehicles, such as engine status, tire pressure, battery status, vehicle speed, and so on, we think the context data of driving, such as road condition, weather, traffic information, also plays an important role in decision making, e.g., assistant driving, battery cell management, remote diagnostics, abnormal driving behavior detection, and so on. Therefore, we propose to include a driving data integrator (a.k.a DDI) in *OpenVDAP*. The main objective of DDI is to automatically collect and store relevant context information on the vehicle and to serve high level services via a set of APIs.

As shown in Fig. 2.5, the DDI consists of three layers. The bottom layer is the data collector layer. The middle layer is the database and the top layer is the service layer. The data of DDI consists of four aspects: vehicle driving data, weather information, traffic condition, as well as social web information like some emergencies. OBD reader and on-board sensors collect the driving data, which includes the location, speed, acceleration, angular velocity, and so on. Noted that, we used an OBD reader since most of the normal vehicles only provide an OBD interface to obtain driving data, and in the future, we will adapt this to more types of vehicles by multi-fold devices, such as CAN card for electric vehicles. Weather, traffic, and social data are collected from vehicle-specific APIs. Environment information gives the weather conditions and real-time traffic conditions, which can be a necessary part for detecting the abnormal driving behavior. Social web information is used to help keep the vehicle away from troubles. In this way, for

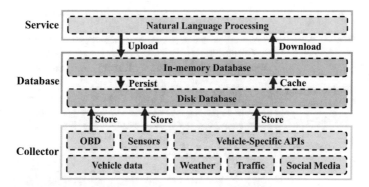

Fig. 2.5 The design of DDI

a specific person driving through a period of time, DDI can provide the driving information, real-time weather, real-time traffic conditions as well as emergencies that have happened nearby.

The database layer is composed of two types of databases, in-memory database, e.g., Redis [433], and disk database, e.g., MySQL [372]. As the data from the collector layer is time-space related, disk database is utilized to store it. Meanwhile, in-memory database caches the frequently used data from disk database to decrease the response latency of request. For all the data caches into the in-memory database, a survival time is set for it. Therefore, all the request for the data would search the in-memory database first, when it cannot be found in in-memory database, it would go to the disk database. All the related data includes location and timestamp. Collected data are permanently stored in the disk database.

The service layer takes charge of requests from the upper layer like `libvdap` via a set of APIs. The requests include two types: download requests and upload requests. An upload request is for users to upload their data onto the DDI while a download request is for the user to download data from DDI. For a download request, the service layer extracts keywords like location and timestamp, then it goes to in-memory or disk database to get data. For an upload request, firstly the data would be stored in in-memory database, when the survival time is up and the related charts have been created in disk database, the data in in-memory database would be written to disk database for data persistence.

Open Problems Here are some open problems that need to be addressed in the design and implementation of DDI. First, what are the right mechanisms to get real-time data, e.g., traffic condition, social web events? It is hard to obtain real-time data with an unreliable, limited bandwidth communication channel available. Second, how to efficiently store and manage time-space related data on a vehicle is challenging. For example, how long will these data need to be stored is still unclear.

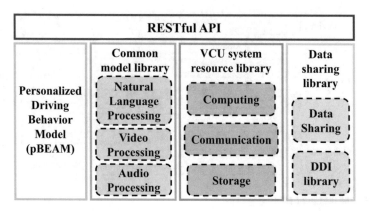

Fig. 2.6 The overview of `libvdap`

2.2.6 `libvdap`: *Library for Open Vehicular Data Analytics*

With the burgeoning of AI-based applications developed for CAVs, *OpenVDAP* should support a variety of artificial intelligence algorithms and models. However, the edge node is not a good fit for executing large-scale models since they require large footprints on both the storage and computing power. To support edge intelligence, `libvdap` is provided which stores many AI common algorithms and models. These models are compressed based on the powerful models and can run smoothly on the edge node. Based on `libvdap`, developers can build AI applications more openly and easily.

As shown in Fig. 2.6, `libvdap` provides a uniform RESTful API. By calling the API, developers can access all software and hardware resources. The resources can be grouped into four categories: Personalized Driving Behavior Model (pBEAM), Common model library (cBEAM), VCU system resources library, and Data sharing library. For example, developers can use the real-time or history data acquired by the Data sharing library, leverage the pBEAM or the AI models provided by the Common model library to build their application. The application will run on the *OpenVDAP* by calling the VCU System resources library. In this way, developers can create applications that provide value in the easiest possible way.

Personalized Driving Behavior Model (pBEAM), the core component of `libvdap`. pBEAM models personalized driving behaviors based on driving data. It has been built and deployed on vehicles. Developers can build third-party applications on top of pBEAM. By leveraging pBEAM, developers can acquire the driver's behavior without a long-running process of collecting and analyzing data. For example, the insurance company can evaluate whether the driver is aggressive or not based on pBEAM.

As shown in Fig. 2.7, pBEAM is pre-trained on the cloud server and trained again on the XEdge (*OpenVDAP*) to obtain the personalized characteristics. On the cloud server, we build a *Common Driving Behavior Model (cBEAM)* based

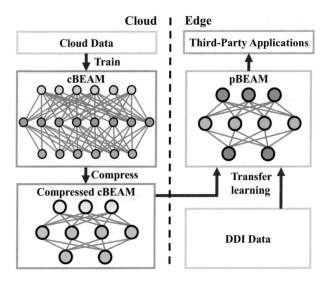

Fig. 2.7 The building process of pBEAM

on a large training dataset which includes many drivers' driving data. The input data includes the location, speed, acceleration, and so on. Although cBEAM is powerful, its scale is large. So it requires large footprint on both the storage and compute power, which is impractical for the XEdge. To save storage and computing resource, a compression algorithm based on Deep Compression [192, 193] is used, in which cBEAM is pruned first to reduce the number of connections by learning only the important connections, then the number of bits for representing each weight is reduced via the weight sharing technique. The compressed cBEAM is then downloaded to the vehicle. Transfer learning [400] is used to transfer the compressed cBEAM to pBEAM by learning the personalized driving data which stores in the DDI. When pBEAM is already trained, it can provide service for third-party applications.

The *common model library* contains many common algorithms and models that are used frequently in vehicle-based applications, such as Natural Language Processing, Video Processing, Audio Processing, and so on. The most powerful models that we leverage today are too large for the *OpenVDAP* to run, so the models that are in the Common model library are compressed based on the powerful models. After compressing, the models are optimized based on the computing power of VCU. The *VCU system resources library* provides developers with the library of system resource. By calling the library, users can access the computing, storage, and communication resources. The *data sharing library* provides to the user the uniform data interface. The data comes from two sources: Data Sharing module of EdgeOS$_v$ and DDI.

Open Problems Several open problems that need to be addressed for libvdap include: Firstly, although we compressed the large-scale artificial intelligence models in the cloud, they are still too large to leverage on the XEdge. Can we design a model which has a smaller size based on personalized dataset? Secondly, how can we consider more context information when build pBEAM, such as driver's age group? At last, the driver may show different behavioral characteristics due to the different kinds of weather. How can we take the environment into consideration when building the pBEAM?

2.3 HydraSpace: Computational Data Storage for Autonomous Vehicles

2.3.1 Introduction

The data captured by an autonomous vehicle is growing quickly, typically generating between 20 TB and 40 TB per day, per vehicle [143]. This includes cameras, which tend to generate 20–60 Mbps, depending on the quality of the image, as well as sonar (10–100 kbps), radar (10 kbps), LiDAR (10–70 Mbps), and GPS (50 kbps) [98, 514]. Storing data securely and efficiently can accelerate overall system performance. Take object detection as an example; the variation of historical data could contribute to the improvement of detection precision using machine learning algorithms. Map generation can also benefit from the stored data in updating traffic and road conditions appropriately. In addition, the sensor data can be utilized for ensuring public security as well as predicting and preventing crime. The biggest challenge is to ensure that sensors are collecting the right data, and it is processed immediately, stored securely, and transferred to other technologies in the chain, such as infrastructure, Road-Side Unit (RSU), and cloud data center. More importantly, how to create hierarchical storage and workflow that enables smooth data accessing and computing is still an open question for the future development of autonomous vehicles.

The current vehicle data logging system is designed for capturing a wide range of signals of traditional CAN bus data, including temperature, brakes, throttle settings, engine, speed, etc. [231]. However, it cannot handle sensor data because both the data type and the amount far exceed its processing capability. Thus, it is urgent to propose efficient data computing and storage methods for both CAN bus and sensed data to assist the development of self-driving techniques. As it refers to the installed sensors, camera and LiDAR are the most commonly used because the camera can show a realistic view of the surrounding environment while LiDAR is able to measure distances with laser lights quickly [493, 550]. Both produce a massive amount of data that would be multiplied with the increased resolution and number of channels.

Many researchers have investigated image compression to save storage spaces [442, 486, 522, 610, 612]. The authors in [425, 590] proposed an efficient image compression algorithm for gray scale images based on the quadtree decomposition method. Other researchers have focused on bit-error aware lossless compression algorithms for color image compression subject to the bit-error rate during transmission [410]. In order to largely reduce the compressed size, lossy compression has been proposed; it does not restore the original data entirely, but the information loss has little impact on the understanding of the original image, resulting in a much larger compression ratio [108, 110, 253]. However, none of these methods has been utilized on the vehicular dataset to reduce the size and save storage space. With more data produced by connected and autonomous vehicles, there is a need to create an effective data store and management plan to facilitate the development of autonomous vehicles and their applications.

In this chapter, we propose a novel computational data storage solution for autonomous vehicles by adopting effective compression algorithms based on different incoming sources. Comprehensive and intensive experiments were conducted to verify the effectiveness of our proposed method. The major contributions of this work are as follows:

- Propose a computational storage architecture named *HydraSpace* to efficiently support a variety of applications leveraging diverse data sources for autonomous vehicles.
- Apply various compression algorithms to investigate the compression performance to find an optimal solution for *HydraSpace*.
- Conduct intensive experiments on an indoor Mobile platform HydraOne [548] to collect a real vehicular dataset and test the system performance as well as the power consumption of multiple sensors.
- Discuss five open questions to envision the future storage design challenge for autonomous vehicles.

2.3.2 Real-time Application Requirement

HydraSpace is a multi-level computational storage system that is designed for autonomous vehicles to compute and manage vehicular data based on access frequency, data type, and volume, as well as real-time application requirements. More importantly, how to support applying multiple machine learning models to the same dataset concurrently also poses a challenge in the design of *HydraSpace*. This calls for creating an intermediate results layer in storage to avoid redundant computations. Figure 2.8 shows the overall architecture of *HydraSpace*.

The storage architecture of *HydraSpace* includes a cache, solid state drive (SSD), and traditional hard disk drives (HDD). This multi-level storage scheme is designed to cater to the vast amount of data generated by multiple sensors, various data access frequency, information backup, as well as data retrieving and analysis. As shown

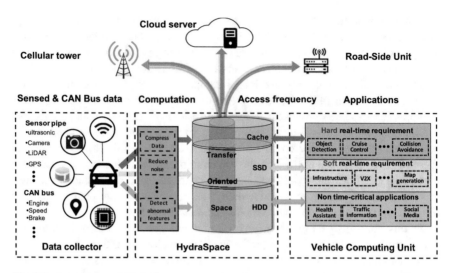

Fig. 2.8 An overview of *HydraSpace*

in Fig. 2.8, there are three types of on-board computing applications that need the support of sensed data. To satisfy the requirements of hard real-time processing applications, such as adaptive cruise control and object detection, the data will be placed directly into a high-speed cache to accelerate the response time. For less time-critical applications that are defined as a soft real-time requirement in Fig. 2.8, the sensor data will adopt a fast lossy compression algorithm and be stored in lower latency SSD to save storage space and meet the quick response requirement. For those non time-critical applications, the data will first be compressed using a lossless algorithm and stored in HDD to satisfy the needs of large capacity while reducing the total cost.

2.3.3   Access Frequency

The sensed data should be placed in different levels of storage architecture based on its access frequencies. The highly frequently utilized data can be arranged into a high-speed cache with limited space; others could be stored in SSDs or HDDs based on their volume and the application demands. For those hard real-time applications, the tasks will be running periodically to generate the results for the vehicle control system to take corresponding action. In other words, these data are considered as "hot" data, which will be accessed frequently by time-critical applications, such as object detection, collision avoidance, adaptive cruise control, etc. As we mentioned in the 2.3.2, the less utilized data can be put into SSD or HDD for backup and later analysis.

2.3.4 Data Amount and Type

There are two major types of data flowing into *HydraSpace* in autonomous vehicles, which are sensor pipe data and CAN bus data. The sensor pipe data contains the camera, LiDAR, millimeter radar, ultrasonic, and GPS. The CAN bus data consists of traditional vehicle data such as speed, engine information, brake, temperature, etc. Compared to the can bus data, sensor pipe data will generate more data due to its complicated data structure and number of different sensors. In this chapter, the data produced by the camera and LiDAR are the major concerns we considered in our design. The detailed analysis of the vehicular data set structure is presented below to demonstrate the in-depth analysis of our proposed solution.

2.3.4.1 Camera Image Dataset

The image data that we used in our paper are RGB images, which are collected using the camera installed on our mobile platform. The RGB image is a three-dimensional array: two of the dimensions specify the location of a pixel within an image, and the other dimension specifies the color of each pixel. The color dimension always has a size of 3 and is composed of the red, green, and blue color channels of the image. We formulate the RGB image using the following equations:

$$\begin{cases} X = var_R * 0.4124 + var_G * 0.3576 + var_B * 0.1805 \\ Y = var_R * 0.2126 + var_G * 0.7152 + var_B * 0.0722 \\ Z = var_R * 0.0193 + var_G * 0.1192 + var_B * 0.9505 \end{cases} \tag{2.1}$$

where X, Y stands for the pixel position within an image, Z represents the specific color of the corresponding pixel, and the parameters represent the specific color map of our selected RGB images.

2.3.4.2 LiDAR Point Cloud Dataset

LiDAR is one of the most popular sensors installed on connected and autonomous vehicles to collect data and support the self-driving applications, such as 3D map generation, and collision avoidance. It will generate a set of points embedded in the 3D space and carry both geometry and attribute information. First, LiDAR acts as an eye of self-driving vehicles. It is continuously rotating and sending thousands of laser pulses to collide with the surrounding objects. An on-board computing system records the resulting light reflections and translates this rapidly updating point cloud into an animated 3D representation [493]. Secondly, LiDAR can be used for state localization. It can obtain precise and real-time locating information in the global coordinate system by matching the online frame with the prior map, which is needed in many modules, such as behavior decision, motion planning, and feedback control. With the help of LiDAR, autonomous vehicles travel smoothly and avoid collisions by detecting obstructions in advance. This improves the safety of commuters and makes autonomous vehicles less prone to accidents due to the risk of human negligence and reckless driving. The downside of LiDAR is that it will produce a huge amount of data up to 30 MB per second (30 MB/s) based on its specific configurations. In this case, managing this data in a fast and effective way could largely contribute to the development of connected and autonomous vehicles. In our proposed *HydraSpace*, we present a sensor pipe data management scheme to abide by the following rules. For hard real-time processing applications, the sensor data can be quickly accessed directly for computing purposes. For applications with soft real-time or fast processing needs, the data can be first stored into the *HydraSpace* and then retrieved by the applications when needed. The point cloud data in 3D space can be translated using the following equation:

$$(X, Y, Z) = H^{-1} * (r, \alpha, \epsilon) \tag{2.2}$$

X, Y, Z stands for the position of each point in 3D space. The object can be described in the following equation:

$$O_s = [O_s^{(1)}, O_s^{(2)}, ..., O_s^{(N)}] \tag{2.3}$$

2.3.5   System Design

Based on the above requirements, we propose a computational storage *HydraSpace* with a multi-layered storage architecture and adopt effective compression algorithms to manage the sensor pipe data. The term "computational storage" means raw data will be stored after applying certain computation processes, such as compression, noise filtering, and abnormal detection. Compared with the previous method of storing raw data, the processed data can effectively improve application performance. Take data compression as an example; the compressed data can largely improve the running efficiency in streaming processing and face model recognition as well as deep model generation [329, 408, 464]. Additionally, the capacity of the cache, SSD, and HDD can be adjusted based on the application requirements and real budget. In our current design, the storage capacity is gradually increased to meet the needs of a large capacity of sensor pipe data and maintain a reasonable cost. To assist in the development of connected and autonomous vehicles and ensure the smooth communication between CAVs, edge servers, and remote clouds, *HydraSpace* will allocate a "transfer oriented space" as shown in Fig. 2.8 to store the temperate data that is ready to be collected by the applications running on edge servers, road-side units, or clouds.

2.3.6 Summary

With the explosive growth of data collected by multiple sensors, data storage is the critical factor for the future of Autonomous Vehicles. Unfortunately, there has been no solid and comprehensive research conducted in this field. Therefore, we propose a computational storage *HydraSpace* with multi-layered storage architecture that adopts effective compression algorithms to manage the sensor pipe data. We also conduct a comprehensive experiment on the indoor experimental platform to study the system performance regarding CPU and memory utilization, power consumption for different sensors, as well as the data set compression with the consideration of different data types. This would be extremely helpful for the application development and system optimization for future autonomous vehicles. For future work, we will focus on the evaluation of various computing platforms, reduce the data stored in *HydraSpace* by incorporating the feature extraction of the collected data, and implement more computation processes. In addition, five open questions and challenges have been discussed to envision the future development of storage for autonomous vehicles.

2.4 AC4AV: A Flexible and Dynamic Access Control Framework for Connected and Autonomous Vehicles

2.4.1 Introduction

With the fast development of sensing, communication, and artificial intelligence technologies, Connected and Autonomous Vehicles (CAVs) have attracted a great deal of attention from industry and academia [71, 317, 556]. Several autonomous driving systems or commercial products have been released in the industry, such as the Google Waymo [554] vehicle, the Tesla Autopilot system, and the Baidu Apollo platform [15]. With the liberation from driving, increasingly more applications, especially various third-party applications envisioned by Zhang et al. [600], will be installed into future CAVs, as supplements to other three kinds of applications, i.e., Advanced Driver-Assistant System (ADAS), real-time diagnostics and in-vehicle infotainment, to enrich the ride experience. Note that some applications are cross-cutting because they fall under more than one category. However, all of them utilize vehicle sensing data, sensed by a plethora of diverse sensors, to realize their functions. For example, the ADAS leverages the data of installed cameras, light detection and ranging (LiDAR), radio detection and ranging (radar), as well as vehicle status captured from Controller Area Network (CAN) to perceive the surroundings and an attack detection application to access the in-vehicle sound data captured by the microphone as the input of its speech recognition [313]. That means the sensed life-critical data is not only used as the input of ADAS, but it is also used by various third-party applications. The malicious applications may preempt

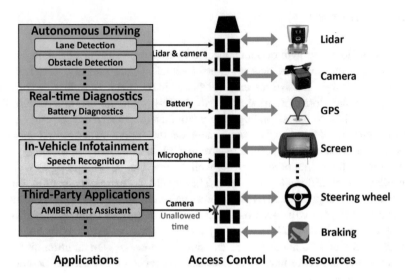

Fig. 2.9 The function of an access control system

the limited computing, memory, storage, and network resources to perform their purpose after obtaining data from the CAV system, which will affect the safety of CAVs. Some malicious third-party might tamper data, leading to wrong decisions on driving, even threatening personal and public safety.

In prior researches, the access control technique [296, 332, 457] is used to protect data from malicious applications by rejecting unauthorized access, which is a security technique regulating who or what can view or use resources in a computing environment. Figure 2.9 illustrates the usage of access control technique in CAV. However, most current researches on CAVs focus on the implementation of autonomous driving vehicle prototypes, including hardware, autonomous driving algorithm [75] and platform [15, 319, 556], and the access control framework enabling the function of applying access control technique to vehicular data is lacking in these researches. Hence, the first thing is to build an access control framework for future CAV. However, the future CAV, with various applications, vehicular data, and different access patterns on different vehicular, is more complex, thus no one existing access control framework could be applied on CAV, directly. But existing mechanisms used by some operating systems [481, 545, 558] provide some experiences for referencing.

Typically, different access control models are suitable for different scenarios with different characteristics [429]. For example, attribute-based access control models with high confidentiality are widely used in a cloud-based storage scenario [450]. However, its computing resource costs are high and it might not be suitable for CAV data. In the CAV area, both performance and security are of high priority. Thus how to choose suitable access control models for different vehicular data is an open problem since no one knows the effects after applying one access control model. But, in any case, some characteristics in the system level should be supported that make the implementation of access control models be more easy and flexible. Firstly, some novel access control models might be proposed for applications, especially for as yet unforeseen applications. Thus, an open access control framework is required for access control research. Secondly, the access control framework should be fine-grained. Taking the permits of the steering wheel as an example, it only could be controlled by ADAS applications, but could be monitored by many other applications. Thirdly, access control should be dynamically changed with the context of CAV and system status. For example, in the application proposed in [313], considering the user privacy, the permit to access inside video should be dynamically gained and revoked depending on the recognition of a "help" signal from an inside squeal voice. Finally, the framework should support applying different access control models to the same data with different grains or different data.

However, the design of access control architecture for the future is challenging as it must fulfill the above requirements while it must meet the intrinsic complexity of vehicular data, including historical and real-time, and access patterns of different vehicular data. Furthermore, there exist various vehicular data and applications in one CAV, resulting in another challenge when developing such an access control framework for future CAV. Generally speaking, an access control framework should know and identify which application is accessing which vehicular data with which access operation. Here naming is a problem, especially for supporting of fine-grained and dynamic access control, an easy-to-read and organized naming mechanism is important so that researcher and user could easily set access permissions for different applications, data, and operations.

To tackle the aforementioned issues, in this chapter, we first introduce the data generated and stored in CAVs and its access patterns based on our observations. Then, we introduce designed access control architecture and data abstraction method to identify application, data, and operation. The proposed framework serves as the access control part of our previous work, Open Vehicular Data Analytics Platform (*OpenVDAP*) [600], which is a full-stack edge supported platform that includes a series of heterogeneous hardware and software solutions. We also implement an access control framework prototype based on the proposed architecture, which responds to queries for access actions and records these actions for future auditing. The contributions are summarized as follows:

- This work is the first to define the data access control problem in CAVs. According to the observations on several CAV platforms, we introduce the characteristics of data and access pattern in emerging CAVs, in terms of real-time data and historical data, while different access patterns are applied. Moreover, different and some new access control models are required.
- We propose a three-layer access control architecture to protect data on CAVs from unauthorized access. The designed architecture supports fine-grained and dynamic access control, and it is extensible with APIs to assign customized

access control models implemented by others, as well as respond to external access actions, resulting in easily extending to meet other access patterns.

- We demonstrate the proposed design through a prototype framework and evaluate it using different access control models. Experiment results show that our framework has a low impact on real-time data and a high but tolerable impact on historical data, which could be solved by periodically caching application information. Furthermore, we also test our framework on an experimental CAV platform implemented based on OpenVDAP architecture, HydraOne [548], which indicates that our framework could run on platforms with different hardware.

2.4.2   Problem Statement

The data in CAVs is important since it affects the decision of autonomous driving algorithms and has implications for passenger privacy. How to protect data from unauthorized access in CAVs is a big challenge. In this section, we first introduce vehicular data access patterns. Then, the requirement of CAV's access control framework supporting different access control models is analyzed. Finally, we formally present the problems of designing an access control framework for future CAVs.

2.4.2.1   Data Access Pattern in CAV

As observed in [603], there are four categories of applications, consisting of ADAS [75, 79], real-time diagnostics [10], in-vehicle infotainment, and third-party application [420, 603]. The data accessed by these applications could be classified into two categories, real-time data and historical data, based on the observation of several CAV platforms, which will be accessed with different access patterns. Thus, we will analyze the data access pattern. Table 2.2 lists the storage locations and potential access patterns of these data.

Table 2.2 Real-time data and historical data in CAVs

Category	Examples	Storage location	Potential access pattern
Real-time	GPS, video, Radar, LiDAR cloud point, engine load, etc.	Pub/sub System	Pub/sub
Historical	GPS, traffic data, and metadata of unstructured data.	Database	Web service
	Video, LiDAR cloud point data.	File system	Web service

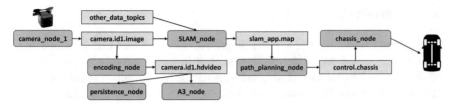

Fig. 2.10 An example data flow of a real-time camera data

Real-Time Data The main requirements for real-time data access are low latency and one-to-many communication since different applications might access the same real-time data at the same time. Most existing CAV solutions utilize normal or modified versions of the Robot Operating System (ROS) [317], which provides the publish/subscribe pattern for different applications of CAVs. Taking Apollo as an example, which is an open-source CAV platform, including hardware reference, system, software, and autonomous driving algorithms, it modified ROS as an underlying system and utilizes message-based communications (publish/subscribe pattern) to deal with one-to-many communication (shared memory technique to reduce the latency of data transmission after version 3.5). In academia, *OpenVDAP* [600] also utilizes a message-based architecture to enable communications of real-time data between devices and applications.

Figure 2.10 illustrates an example data processing flow of real-time camera data under the publish/subscribe pattern. The camera pushes raw images to the topic of *camera.id1.image*, and several processing nodes subscribe this topic while the encoding node encodes images into videos for persisting camera data into the file system as historical video data, and the SLAM node analyzes images for autonomous driving. The path planning node (also an autonomous driving application) subscribes the output of the SLAM node and publishes the control data for chassis control.

Historical Data For historical data, current CAV solutions persist real-time data using the ROS built-in function, which directly saves data as ROS packages. However, it is not a good way for future CAVs. A simple way is storing structured data (e.g., GPS data) into a database and unstructured data (e.g., video) in the file system. In this case, the application could inquire about the structured data from the database directly, or inquire about the storage path of the unstructured data from the database; then it could access the file in the file system.

However, it is insecure to provide a database interface for a CAV on the road. Thus, a centralized manager is needed. To this end, Zhang et al. proposed a module, *Driving Data Integrator (DDI)* in the *OpenVDAP* platform [314, 600], to automatically collect and store relevant context information on the vehicle and the Internet. The application could inquire data from this service. In addition, we need to note that the unstructured data could be accessed through the file system (paths are queried from DDI) or through the DDI service as a more secure method.

2.4.2.2 Access Control

The access control technique aims to protect data and resource in a computer system from unauthorized access. Typically, it includes several concepts, such as access control framework and access control model. The former one captures access actions in an application system or operating system and applies one of access control models to authenticate the access actions. As mentioned before, different types and access patterns of data exist in CAV, and different applications are willing to manage their data in different ways. Specially, for these as yet unforeseen applications, some novel access control models might be proposed.

Based on the characteristics and requirements of the applied scenario, various access control models have been proposed, such as Role-Based Access Control (RBAC), Identify-Based Access Control (IBAC), and Attribute-Based Access Control (ABAC) [112]. For example, the historical battery information under IBAC model could be shared with the ones who have the identity certificate issued by the car maker. Meanwhile, access control models founded upon fine-grained and attribute-based encryption could secure data and prevent unauthorized access (without right attributes), which we will introduce in Sect. 2.4.4. Thus, how to support several access control models thus provide a flexible and suitable choice for CAV and CAV application developers is still a big challenge.

Moreover, the context of CAV is also important for access control. For instance, third-party applications are prohibited from accessing network resources due to insufficient network bandwidth. Or, third-party applications are prohibited from accessing camera resources to avoid privacy leak, when the CAV is in a special location.

Problem Statement Current access control frameworks usually focus on only one type of data. For example, access control frameworks in the operating system, messaging system, and web service system focus on access of file systems, topics, and HTTP requests, respectively. Thus, to protect data with a suitable access control models in CAVs with various applications, data, data access patterns, a systematical and flexible access control framework is needed, which is enable to face possible changes occurred in the future, firstly. Here several barriers must be solved as follows: (1) how to design a framework to authenticate these access actions from different data sources with different access patterns? (2) how to enable the supporting different access control models in one access control framework, and also enable the development of new access control models? (3) how to uniquely identify various data in access actions and access control models from different data sources? and (4) how to dynamically make decisions on access actions based on current vehicle status, including location, computing resource, network resource, as well as supporting different access control models for different data?

2.4.3 System Architecture

We have introduced the motivations and goals for access control framework in future CAVs, and now we will present our design. First, we will introduce some concepts in an access control framework, followed by the security and threat model. Then, we primarily focus on introducing the proposed access control framework for future CAVs.

2.4.3.1 Definition

A traditional access control framework will authenticate one `subject` whether it has the permission of one type of `operation` to one `object`. Thus, an access action could be described as a tuple {`subject, object, operation`}. In detail, the descriptions of `object`, `subject` and `operation` in our CAV-specific framework are as follows.

Subject Various applications, including native applications (e.g., ADAS applications) and third-party applications, are installed on CAVs, enhancing the ride experience and public safety, and they need to access and analyze data sensed by CAVs, thus becoming `subjects`.

Object The `objects` refer to the data in CAVs, e.g., real-time data and historical data. It is easy to understand, and most applications analyze sensed data (as `objects`) for autonomous vehicles. Additionally, remote data, such as road conditions or weather data from remote cloud servers or other vehicles also, is included. Furthermore, the application data is also the `object` in CAVs, since some applications might share/require results from collaborative applications.

Operation As mentioned above, there exist several data sources in CAVs, including publish/subscribe system for real-time data, Web-based service for structured data and file system for unstructured data. Thus, the operations defined in access actions currently include `subscribe/publish`, `get/post/delete` and `create/read/write/delete`, respectively.

In this chapter, we aim to build an access control framework supporting different access control models with dynamic adjustment. Thus, we add a segment `extra` to that tuple, which is used to store additional information. In the following sections, the tuple is defined as follows.

$$\{subject, object, operation, extra\} \qquad (2.4)$$

Fig. 2.11 The architecture of AC4AV

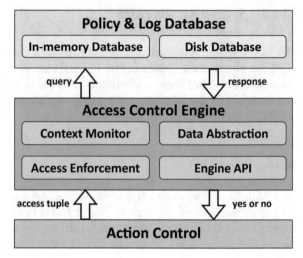

2.4.3.2   Security and Threat Model

The malicious applications in our threat model always try to subscribe to the data-related topics in the Pub/Sub system, query historical data from the database or file system. Typically, applications are isolated utilizing container technology, and they cannot access others' memory to obtain data, such as subscribed data and the authentication information. In this chapter, we do not consider the leakage of authentication information and it could be secured by other approaches, such as secure storage. Additionally, the trusted execution environment is also promising to provide the isolation of applications, by executing part of one application in a hardware-assistant environment, so that the running application can be protected from not only other applications but also the operating system and even hypervisor. Moreover, our AC4AV also can be executed in that trusted execution environment, such as Intel Software Guard eXtensions (SGX) or AMD Memory Encryption Technology [358], which could significantly reduce the attack surface.

2.4.3.3   Architecture of AC4AV

The proposed access control architecture for CAVs is as shown in Fig. 2.11, which consists of three components, *Access Control Engine*, *Action Control* and *Policy & Log Database*. The *Access Control Engine* authenticates operations and responds yes or no to the *Action Control* component, which performs as a hook to capture access actions. The *Policy & Log Database* stores all data of AC4AV, such as the configuration file and the access action record, in a hierarchical mode while frequently used data is stored in the in-memory database with high-speed access.

Access Control Engine In the *Access Control Engine*, we introduce four major components: (1) *access enforcement*; (2) *context monitor*; (3) *data abstraction*; and (4) *engine API*.

Since the goals of our access control system include dynamic access authentication, the *context monitor* component is used to collect the system's status information, such as CPU utilization and GPS, enabling an access control policy with dynamic features. For example, the A3 application would gain the access permissions of the historical video data with the specific location and time ranges.

As discussed in Sect. 2.4.2, real-time data are different from historical data in terms of their access patterns, and different data identifying methods are applied by different systems. Thus, we designed the *data abstraction* component, which provides the function of identity conversion. It converts the identities of `objects` and `subjects` in captured tuples from different data sources and the internal identities to each other, based on a data abstraction method. Benefiting from this component, third-party applications also could identify their own data using an easy-to-read description method when they implement customized access control policies. The data abstraction method is introduced in Sect. 2.4.4.

To improve the expandability of our AC4AV, we designed a component, named *Engine API*, with a series of APIs. The provided functions are multi-fold: (1) to support a customized policy, a series of APIs for customized policy configuration are provided for applications to a submitted policy file. Moreover, several implemented access control models are provided for AC4AV so that applications could assign different access control models to protect their data; (2) to support auditing, a series of APIs are provided for inquiring records of access actions, as well as results; and (3) to configure AC4AV, a series of APIs are provided so that the system administrator could configure all components in AC4AV . For example, we will not limit the database used by *Policy & Log Database*, so the system administrator could assign it in the configuration file or adjust it through the provided APIs in runtime.

The last component of the *Access Control Engine* is the *access enforcement*, which is a core component, just like an assembler, combining other components to determine the permission to access actions. While a `subject` intends to access the `object` with the `operation`, all the related information of that access action, as shown in tuple {`subject`, `object`, `operation`, `extra`}, will be captured and sent to this component. Then, it will send the segment of the `object` to the *data abstraction* to figure out the internal identification of the accessed data. Then, it will authenticate this action using the specified policy, associated with other factors, defined in the policy and collected by the *context monitor* component. The output of this component indicates whether the `subject` has corresponding permissions.

Action Control Service To capture access action and deny the action without permissions in different data sources, e.g., message queue system and web-based service, the *Action Control* service is proposed in our AC4AV, mainly implementing two functions, *action capturing* and *action responding*.

The *action capturing* function is that implemented sub-service captures all access actions and submits these actions to the *Access Control Engine* with the format as the tuple {subject,object,operation,extra}. The *action responding* refers to sub-service performing corresponding operations (allowing or denying) based on the responses (yes or no) from the *Access Control Engine*. Note that the ways for allowing/denying operations will vary depending on the corresponding access method. For instance, a NGINX module will be implemented for capturing all data access queries to DDI module in *OpenVDAP*, and it will reject all non-permission queries with a 403 status code if it receives a no from the *Access Control Engine*. Thus, the *Action Control* service will consist of various sub-services when implemented.

Policy and Log Database Vast access action records should be recorded for future auditing, and many policy files should be collected in AC4AV . Thus, a *Policy & Log Database* is proposed to store this data. It is a two-layer architecture consisting of two database systems.

The lower layer database is a *disk database*, and it stores all information of AC4AV, including access action records, policy files, configuration files, and so on. However, the *disk database* is with a high query latency, and typically, one access control operation requires a quick response. An *in-memory database* is proposed as the upper layer database to reduce the inquiring latencies of these frequently used policies. Once the AC4AV starts, the *Access Control Engine* will inquire those policies from the *disk database* and then store the parsed policies to the *in-memory database*. We should note that all the stored data is only allowed to be accessed by the *Access Control Engine*. The reason is that stored data, i.e., access action records, policy files, configurations of AC4AV, is important, especially when auditing in an accident investigation.

2.4.4 Implementation

In this section, we introduce the implementation of our AC4AV prototype. First, we introduce the data abstraction method. Second, the prototype implementation of AC4AV is illustrated, followed by three access control models which are implemented as built-in models in the *engine API* component.

2.4.4.1 Data Abstraction Method

The naming of objects from different data sources are different; thus, how to identify subject and object is a problem that must be solved in an access control framework. To this end, an easy-to-read data abstraction method is proposed when implementing the prototype of our AC4AV, which converts the identities of `objects` and `subjects` in the captured tuples from different data sources and internal identities to each other.

Object Different data sources have their own methods to identify data. The following equation is an example identity on the Pub/Sub system for the installed front camera, which publishes video data for applications. Thus, one application could obtain real-time video by subscripting such topic.

$$camera.id1.channel\_720P \tag{2.5}$$

In addition, historical data could be accessed through the web-based service. Taking the historical video data of the foregoing camera as an example, one application could request that data with a specific time period (e.g., $t1$ to $t2$) from the DDI service. The access URL is as listed in Eq. 2.6. Note that we omitted "`http://`" in the equation due to limited space. In this case, the identity of the historical data is different from the real-time data, shown in Eq. 2.5.

$$ddi/camera?type = id1\&start = t1\&end = t2 \tag{2.6}$$

To uniformly identify the data in our `AC4AV`, we leveraged Uniform Resource Identifier (URI), which is a hierarchical method. Typically, the `object` can be described in three parts: owner, identifier, and parameters. The identities listed in Eqs. 2.5 and 2.6 can be, respectively, described as following based on our method.

$$/sys/camera/front/realtime?resolution = 720 \tag{2.7}$$

$$/sys/camera/front/history?start = t1\&end = t2 \tag{2.8}$$

Here, `/sys` indicates the owner, the identifier of data is `camera/front/realtime`, and `resolution=720` assigns the resolution of the subscribed video stream, while a real-time video source might publish several streams to the system with different resolutions. Similarly, the historical data sensed by the same sensor can be identified by replacing `realtime` with `history` and setting some parameters, such as start and end timestamps. Based on this method, the `object` can be uniformly identified in the `data abstraction` component, regardless of the `object` in the publish/subscribe system or in the DDI service.

Subject In addition, we also use this method to identify `subjects` (applications). For example, the application A3 [603] could be defined as `/com/qyzhang/A3?v=1.0`, while `com/qyzhang` is the owner and `A3` is the identifier. Moreover, a group label can be defined to identify a set of applications thus providing a simpler way to manage applications in an access control policy. For example, `group:autonomous-services` is used to refer all services related to autonomous driving. Note that this group label should be created as a system-level configuration through the *engine API* component.

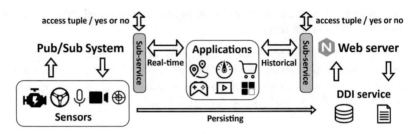

Fig. 2.12 Implemented sub-services for the *Action Control* service

2.4.4.2   Implementation

We implement a prototype of AC4AV based on proposed architecture. Most parts of the implemented prototype are programmed using Golang language, except the *Action Control* sub-service on NGINX and the attribute-based encryption (ABE) algorithm, which is implemented using C/C++ language. In addition, we have tested our prototype on two different platforms. One is a normal desktop with the Intel CPU, and another is our HydraOne platform [548] with the NVIDIA Jetson TX2 processor, which is an indoor experimental research and education platform for CAVs based on OpenVDAP architecture [600].

Action Control
The *Action Control* service consists of various sub-services, implemented to capture all access actions from different systems. In this prototype, we implement two sub-services. One is embedded in a publish/subscribe system for real-time data and another is embedded in a web server for historical data, as shown in Fig. 2.12.

The chosen publish/subscribe system for our AC4AV prototype is NATS [376], which is a simple, high-performance open-source messaging system that provides multi-language clients, such as Python, Java, C, etc. When re-compiled NATS receives a subscription request, the implemented module will capture all information about this action and send this information to the *Access Control Engine* for authentication. If the response is no, it will reject this subscription action. In our prototype, the identity of a subject is captured according to the socket's port used by the application once the application connects to the NATS. The object is the subscribing or un-subscribing topic. In addition, we also modify the protocols of NATS so that the applications can attach extra information in PUB, SUB, and UNSUB protocols for authenticating under different access control models.

Applications could inquire the structured data and metadata of unstructured data from the DDI service, hosted on NGINX [384] in our prototype, which is a popular and high-performance tool to implement a web server. Benefiting from the expansibility of NGINX, we implemented a sub-service of the *Action Control* service using the C++ language, registered to NGX_HTTP_ACCESS_PHASE on NGINX. Similar to the sub-service in the re-compiled NATS, the implemented NGINX module will collect the access information from an HTTP request received

```
/* Update configuration of access control system */
func (*EAPIs) UpdateConf(key string, value string) error
/* Search a policy */
func (*EAPIs) InquirePolicy(data string) (Policy,error)
/* Create a policy */
func (*EAPIs) CreatePolicy(config_json string) error
/* Update a policy */
func (*EAPIs) UpdatePolicy(config_json string) error
/* Delete a policy */
func (*EAPIs) DeletePolicy(config_json string) error
/* Inquire logs */
func (*EAPIs) InquireActionLogList(query string) string
```

Listing 2.1   Interfaces implemented in the prototype

by NGINX to the DDI service. Then, the information is sent to the *Access Control Engine*, and the HTTP query is rejected with a 403 Forbidden HTTP status code if the response is no.

Access Control Engine For the *Access Control Engine*, it is implemented as a RESTful web service. The principle of the *data abstraction* component has been described in the previous subsection. We implemented this component and provide internal functions for other components, i.e., *access enforcement* component. Note that we use an external storage system, i.e., Redis, to store mapping relationships for identity conversion, in our prototype, which causes extra communications on inquiring identities. The reason is to improve the scalability, considering the increase in the number of mappings. For the *context monitor* component, it periodically obtains and caches system running status from underlying system interfaces, e.g., CPU utilization, as well as some privacy information about vehicle status from the modified NATS, such as GPS data. The *engine API* component allows the system administrator to configure the access control framework and third-party applications to update their configuration. Listing 2.1 illustrates the interfaces we implemented in our prototype. The UpdateConfigure interface allows the system administrator to update system's parameter, and the InquirePolicy interface allows the system administrator or third-party application to obtain policies corresponding to the data. Note that we allow the system administrator to view all policies and the application to view its own data's policies. To support auditing, we also implemented the InquireActionLogList interface for inquiring access action records with a set of query conditions. In addition, the remaining three interfaces allow for the management of policies. Similarly, we only allow an application to manage its own policies. Moreover, to reduce the attack surface, only the necessary APIs are open to management. Furthermore, some authentication approaches should be considered when calling these APIs. For example, as mentioned before, a hardware-assisted trusted execution environment could ensure that application codes are not modified and provide local attestation. Thus, secure channels between our AC4AV and other applications may be established [340].

The *access enforcement* component implements the function of the receiving access actions, authenticating permission and responding to the access actions.

```
 1 { "version": 1.0,
 2     "data": "camera/front/realtime?fps=25",
 3     "owner": "system",
 4     "allow": [
 5       {   "operation": "read",
 6           "parameters": {
 7               "encode": "h264",
 8               "width": 1920,
 9               "height": 1080 },
10           "access_model": {
11               "type": "external",
12               "service": "https://localhost:9999/auth"}
13       },{ "operation": "read",
14           "parameters": {
15               "encode": "h264",
16               "width": 1280,
17               "height": 720 },
18           "limit": [ "CPU":"max 0.5" ],
19           "access_model": { "type": "ABAC"}
20       },{ "operation": "read",
21           "parameters": {
22               "encode": "raw",
23               "width": 3840,
24               "height": 2160 },
25           "access_model":{
26               "type": "ACCL",
27               "applications": ["group:autonomous"] }
28       }]
29 }
```

Listing 2.2 An example of access control policy

To describe a policy, JavaScript Object Notation (JSON) is used to represent the policy's features, which is a lightweight data-interchange format. Listing 2.2 shows an example policy. The parameter segment is used to describe the data. Here, we assign three access control models to data with different parameters of resolution, while autonomous driving applications could access 4K video data based on the defined access control model. Note that the h264-encoded video is the output of the encoding_node in Fig. 2.10 (camera.id1.hdvideo) and the raw one is the output of the camera_node_1 (camera.id1.image), while we assume camera_node_1 and encoding_node are the system services for the camera 1. The access control models will be introduced in the next subsection. The limit segment determines the limitations of data on the vehicle status, i.e., CPU, memory, network. Moreover, we can move the same parameter to the data segment, like the "fps=25". In addition, the data owner could authenticate access actions on its own data, by setting type to "external" in the access_model segment. Typically, the external component must be implemented with an interface to accept the tuple and can be used to achieve dynamic access control. Thus the *access enforcement* component will forward the request to the assigned external link defined in the service segment.

Policy and Log Database We implemented this service based on two types of databases. For the disk database, MongoDB [360] is used to store all data, including policies, access action records and configurations. Every access action with the result and related parameters (such as used system status) is recorded by the *Access Control Engine* and stored in MongoDB. The used in-memory database in our prototype is Redis [433], which is an open-source database and provides key-value storage. Redis will cache these frequently used policies, such as the policies of parts of real-time data and created by running applications, and use the `data` segment in JSON files as keys. Further, as mentioned above, the identity mapping relationships are also cached in Redis, while those are also persisted on the disk database.

2.4.4.3   Access Control Model

In this section, we illustrate several implemented access control models used for different applied scenarios in this chapter: Access Control Capability List (ACCL) -based Discretionary Access Control (DAC) model, Identity-Based Access Control model (IBAC), and Attribute-Based Access Control model. And the tuple captured by our `AC4AV` could be used to develop a new access control model.

ACCL-Based DAC Model   An ACCL-based DAC model restricts the access to the regulated `objects` through access control capability lists, which define the set of `subjects` and their `operation` permissions. Here the ACCL-based access control model is a simple DAC model. The model applied to 3840 × 2160 video data in Listing 2.2 is an example of the ACCL-based DAC model. Once the *access enforcement* component receives an access action with the `object` assigned to ACCL model, it checks whether the `subject` is included in the list defined by the `applications` segment. This model is appropriate for the scenario that has clearly known all `subjects` who will access the data so that it could list exhaustively in the policy.

Identity-Based Access Control Model   An IBAC model restricts access to the regulated `objects` through the identity of the `subjects`, which has several ways to implement. In our prototype, we use a certificate-based approach. The data requester (`object`) should obtain the certificate issued by the data owner using a private key and attaching the certificate when inquiring the data. Only the certificate is valid, the `subjects` could access the data. This model is flexible when the owner could make allowed `objects` be part of the issued certificate. Thus, the applications developed by the same company with different application names could access the data without multiple authorizations.

Attribute-Based Access Control Model   In our prototype, an ABAC model restricts access to the regulated `objects` through encrypting the data of the `subjects`. Thus, the data can be public for all applications but encrypted by related attributes, defined in an access structure. Only the data requesters, who have the related attributes, can decrypt the data from ciphertexts. Once the *access enforcement* component receives an access action assigned to the ABAC model, it directly allows the request. Figure 2.13 illustrates an example of an attribute-based access structure for real-time video data captured by on-board cameras. It

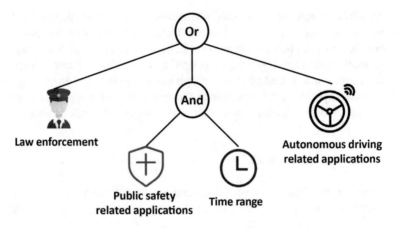

Fig. 2.13 An example of access structure

determines that the one with the attribute of "law enforcement" or the attribute of "autonomous driving applications" could access real-time video data, as well as the one which has the attribute of "public safety application" could access the data in a special time range.

2.4.5 Summary

In this chapter, we investigated the characteristics of data and access patterns, as well as the difficulties of designing and implementing an access control framework in the CAV scenario. To tackle these problems, we designed and implemented a three-layer access control framework to authenticate access actions of real-time data and historical data on different systems, which supports fine-grained and flexible access control models and is extensible with several APIs, enabling configuring access control policy to application and implementing customized access control models. Then, we implemented a prototype that could capture real-time data access actions on the publish/subscribe system and historical data access actions on the web service. In addition, three access control models are implemented as built-in models, and third-party developers could utilize those directly or apply their own access control models through APIs. Finally, we demonstrated our framework by evaluating the performances in the cases of applying different access control models to vehicle sensing data. The results show that our framework has a tolerable impact on access actions.

The current version of our framework has low overhead for access actions, and several improvements could be performed in future studies to further reduce the latency. We will design caching modules for different framework components. For instance, frequently used policies and identity mapping relationships could be cached in the *Access Control Engine* component thus no Redis access is requested. Moreover, we will try to analyze the requirements of access control models in CAVs and design suitable access control models. Furthermore, benefiting from the expandability of our AC4AV, we will implement more *Action Control* sub-services to face changes in future access patterns, e.g., sharing memory mode for real-time data.

Chapter 3
Algorithm Deployment Optimization

3.1 CLONE: Collaborative Learning on the Edges

3.1.1 Introduction

The wide deployment of 4G/5G has enabled connected vehicles and Internet Protocol cameras (IP cameras) as the perfect edge computing platforms for a plethora of new intelligent services. In the meanwhile, with the burgeoning growth of the Internet of Everything (IoE), the amount of data generated by these edge devices has increased dramatically [466]. For instance, a connected and autonomous vehicle (CAV) could produce around one gigabyte of data per second [349] and generate more than 11 TB of privacy-sensitive data on a daily basis [327, 449]. Besides, IP cameras could generate over 2500 petabytes of data per day [147, 475]. Such huge volumes of data inevitably bring challenges to researchers and domain experts—processing big data often requires a large amount of computation and memory resources, which hinders the application of deep learning algorithms on the resource-constrained edge devices with the stringent latency [528].

In this context, cloud-only approaches [56, 240, 325] and cloud-edge methodologies [6, 88, 187, 244] have been proposed to process big data generated by edges in the past few years through computation offloading [6, 582], service scheduling [227, 286], virtual machine migration [237, 334], and so on. These methods both incorporate powerful cloud to resolve bandwidth limitations and provide near real-time services, but they both require sending amounts of raw data to the cloud via the wireless network. Consequently, data transferring may become the latency bottleneck and incur a high bandwidth cost [540]. Besides, even if data is compressed in the edge before being sent to the cloud [187], the original sensitive data might be exposed, which may create a potential threat of privacy leakage. Therefore, it can be seen that various intelligent applications on the edges are facing challenges in decreasing latency and protecting privacy [599].

© The Author(s), under exclusive license to Springer Nature Switzerland AG 2021   57
W. Shi, L. Liu, *Computing Systems for Autonomous Driving*,
https://doi.org/10.1007/978-3-030-81564-6_3

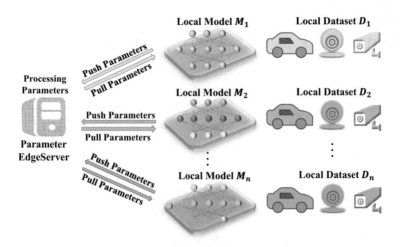

Fig. 3.1 The framework of CLONE. Each edge node trains/runs the neural network model locally based on its private data and push their own parameters to the Parameter EdgeServer during the training/inference process. The Parameter EdgeServer is responsible for performing aggregations or other necessary operations on the uploaded parameters and sending the updated parameters back to the edge nodes

In the meanwhile, as the computation and memory resources of edge devices have become more and more powerful [244], questions arise whether always incorporating cloud to process data is desirable moving forward, and whether we are able to implement novel approaches on the edge side to face the challenges of analyzing big data. With these insights, we propose a collaborative learning setting on the edges (CLONE) which is able to mainly demonstrate the effectiveness of latency reduction and privacy-preserving. The basic framework of the CLONE is shown in Fig. 3.1.

Our core contributions are not in the development of machine learning-based models that are built on top of well-understood and mature models such as long short-term memory networks (LSTMs) [326, 405] and gradient boosting decision tree (GBDT) [256]. Instead, the core innovation of our study is in (*i*) combining state-of-the-art algorithms on top of the Federated Learning concept so that these algorithms can support dynamic distributed learning (*e.g.,* collaborative reliability analysis and computer vision computation), and (*ii*) providing experimental evidence to establish that CLONE could be employed in both model training and inference phases, which provides actionable insights on using the CLONE framework to support other real-world intelligent services that require the collaboration of diverse edge devices.

To be concrete, this chapter presents a collaborative learning framework for edges, which could be used in the training phases (CLONE_*training*) and inference phases (CLONE_*inference*) with the effectiveness of privacy-preserving and latency reduction. Regarding CLONE_*training*, we choose the failure prediction of electric vehicle (EV) battery and related components as the first use case. As to CLONE_*inference*, we choose customer tracking in a grocery store as the second use case, showing that CLONE is a powerful solution to the multi-target multi-camera tracking problem. Specific contributions are listed as follows.

- We have demonstrated the applicability of CLONE in two typical edge computing scenarios, covering both the training and inference at the edges in a collaborative fashion.
- To the best of our knowledge, this is the first work to predict an imminent failure of EV battery and associated accessories based on the real-world EV dataset which involves driver behavior metrics.
- Our analysis reveals that adding driver behavior metrics is able to improve the prediction accuracy for the failures of EV battery and associated accessories.
- CLONE_*training* has the capability to reduce the training time significantly without sacrificing prediction accuracy.
- CLONE_*inference* is a powerful solution to the multi-target multi-camera tracking problem.

3.1.2 System Design

Based on the progress of developing a neural network model, we divide two categories of application scenarios for CLONE, *i.e.,* CLONE in the training stage (CLONE_*training*) and CLONE in the inference stage (CLONE_*inference*). In this section, we introduce the core ideas of CLONE and illustrate the differences between CLONE_*training* and CLONE_*inference*. Then we describe the hardware configuration information for the experiment setup of the two case studies.

Figure 3.1 presents the basic framework of CLONE. These two types of CLONE share the same core ideas—the training/inference tasks are both solved by a group of distributed participating edge nodes which are coordinated by a Parameter EdgeServer. Each edge node has its local training dataset that is never uploaded to the Parameter EdgeServer or transferred to the cloud. In CLONE, each edge node trains/runs the neural network model locally based on its private data and push their own parameters to the Parameter EdgeServer during the training/inference process. The Parameter EdgeServer is responsible for performing aggregations or other necessary operations on the uploaded parameters and sending the updated parameters back to the edge nodes.

3.1.3 Differences of Two Application Scenarios

However, there are three main differences between CLONE_*training* and CLONE_*inference*, *i.e.,* the types of transmitted parameters, the tasks of the Param-

Table 3.1 Two types of application scenarios for CLONE

	CLONE_*training*	CLONE_*inference*
Use Cases	Predict the failures of EV battery and related accessories.	Customer tracking in a grocery store.
Parameters	The value of network parameters.	Appearance descriptors.
Tasks of EdgeServers	Parameter aggregation.	Store, update, and delete trackers.
Transmission	Asynchronous.	Synchronous.

eter EdgeServer, and the transmission manners are different. Table 3.1 summarizes the main differences between CLONE_*training* and CLONE_*inference*, and it serves as a roadmap for the next sections which describe two use cases related to these two application scenarios.

3.1.4   Use Case: Failure Prediction of EV Battery and Related Accessories

In this section, we present our experimental results of CLONE_*training* setting on the first use case—failure prediction of EV battery and associated accessories. The main idea is to predict upcoming failures in a collaborative fashion by capturing individuals' driving characteristics. We also compare the performance of stand-along learning (ALONE) with CLONE_*training* in two aspects: (i) training time, and (ii) evaluation scores including precision, recall, accuracy, and F-measure. After that, we discuss critical observations based on the comparison results.

3.1.4.1   Background

Recently, EVs have received significant attention as an essential part of the efficient and sustainable transportation system. As the key component of EVs, the battery system largely determines the safety and durability of EVs [33, 569]. Due to the aging process or abuse maneuvers during the real operation, various faults may occur at each constituent cell or the associated accessories. Therefore, it is essential to develop early failure detection techniques to ensure the availability and safety of EVs through anticipated replacements. Besides, we believe driver behavior metrics such as speed and acceleration reflect the usage of an EV, which could be the main root causes of EV's failure. Hence, how to build a personalized model to predict the failure of EV battery and related components, and what is the correlation between driver behavior metrics with the failure prediction are both open problems. In this work, we choose the failure of EV battery and related accessories as our case study to show how the CLONE solution trains the failure prediction models to ensure the sustainable and reliable driving in a collaborative fashion.

3.1.4.2 Data Description

The study of the first use case presents an analysis of EV health characteristics based on the data measured at and collected from a large EV company. We analyze three different models of EVs, and the corresponding data is reported and collected every 10 ms during the whole 6-h duration of the data collection period. The battery cells of these three EVs are made by the same battery vendor (BAK company) with different battery types—Lithium Cobalt Oxide (LCO) and Nickel Cobalt Manganese (NCM). The battery pack of each EV consists of 96 battery cells with 30 temperature sensors to detect and report cell temperatures in real-time.

In general, our dataset is collected from the core control systems of EVs, which includes the vehicle control unit (VCU) [335], motor control unit (MCU) [298], and battery management system (BMS) [83]. BMS [83] is responsible for the battery maintenance and state estimation. The VCU [335], as a key component of the whole EV, sends orders to other modules based on the driver manipulation (such as gear signal, accelerator pedal signal, and vehicle mode) via CAN communication network [427]. MCU [298] controls the wheel motor locally according to the command from VCU. Failures in the MCU may cause abnormal motor torque outputs, which in turn affect the vehicle safe driving. Therefore, the functional safety of these systems is particularly important. Figure 3.2 shows the structure diagram of the core control systems.

More specifically, we analyze EV data in the two aspects: (1) EIC attributes, and (2) driver behavior metrics. Here, EIC refers to electric, instrumentation, and computer control systems [294]. It includes battery features collected from BMS (most commonly used for EV battery durability analysis by other studies [33, 148, 409, 459, 575]) and the data reported from other control systems. The number of available features is more than 250, but not all features have useful value (except for NONE and constant value). For our study, 42 features listed in Tables 3.2 and 3.3 were selected as three EVs reported these attributes and the value of these attributes varies over time.

Most of the selected features could be understood intuitively from Tables 3.2 and 3.3; hence, we choose some vague features to give our explanations. Taking account into the CAN communication demand of different modules, the communication

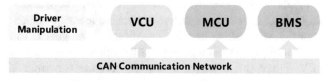

Fig. 3.2 Three core control systems of EVs

Table 3.2 Selected EIC features

Voltage	Temperature	Power & Energy
BMS_BattVolt	InCar_Temp	BMS_BattSOC
BMS_CellVoltMax	Environment_Temp	BMS_MaxChgPwrCont
BMS_CellVoltMax_Num	BMS_BattTempAvg	BMS_MaxChgPwrPeak
BMS_CellVoltMin	BMS_Inlet_WaterTemp	BMS_MaxDchgPwrCont
BMS_CellVoltMin_Num	BMS_Outlet_WaterTemp	BMS_MaxDchgPwrPeak
MCUF_Volt	BMS_MaxTemp	VCU_Batt_Comp_Pwr
MCUR_Volt	BMS_MinTemp	VCU_Batt_PTC_Pwr
Current	BMS_TempMaxNum	Error Info
BMS_BattCurr	BMS_TempMinNum	BMS_BatterySysFaultLevel
MCUF_Curr	MCUF_Temp	BMS_Low_SOC
MCUR_Curr	MCUR_Temp	VCU_PTC_ErrSta

Table 3.3 Selected driver behavior metrics

Driver behavior	VehicleSpeed	Acceleration	Steering
	YawRate	WheelSpeedFL	WheelSpeedFR
	WheelSpeedRL	WheelSpeedRR	EmergencyStop
	AccPedalPosition	BrakePedalPosition	

Driver behavior metrics are collected from VCU and sensors

network has been set several bus nodes, including MCU for the front wheels (MCUF) and rear wheels (MCUR). "MCUF_Volt" and "MCUR_Volt" represent the voltage value of MCUF and MCUR. Positive Temperature Coefficient heater (PTC) is the heating unit in EV. "VCU_PTC_ErrSta" shows the failure status of battery PTC, which preheat the battery at low temperature to make sure the battery is able to work properly. "BMS_BatterySysFaultLevel" indicates the healthy states of BMS. State of Charge (SOC) is the indicator of left capacity, and "BMS_Low _SOC" shows the SOC states of EVs. "YawRate" reflects the overall tilt state of EVs. "AccPedalPosition" and "BrakePedalPossition" show the real-time percent of the gas pedal and brake pedal that the driver pressed on. As to the three error info attributes, they both show the degradation process from healthy to failed. Besides, "Comp" is an acronym of the compressor.

3.1.4.3 Experiment Design of ALONE

Experiment Goals Before employing CLONE, we first combine the whole real-world dataset of three EVs to train different machine learning models on an Intel FRD. We term this stand-alone learning as ALONE. The goal is to figure out a suitable algorithm to predict failures, explore the influence of the driver behavior metrics on EV failure prediction, and provide baseline experimental results to compare it with the algorithm performance of CLONE_*training*.

Experiment Groups To show the impact of driver behavior metrics on the battery failure prediction, we conduct experiments on two experimental groups. Our first step is to combine all selected EIC attributes and driver behavior metrics to train models using different methods, and we label this group as ED Group. Then, we exclude all driver behavior metrics but keep EIC attributes, and we denote it as E Group.

Experiment Setup We tackle the failure prediction problem using random forest (RF) [303], gradient boosted decision tree (GBDT) [151, 581], and long short-term memory networks (LSTMs) [118, 214] since they have become highly successful learning models for both classification and regression problems. The models learned in this use case are implemented in Python, using TensorFlow 1.5.0 [1], Keras 2.1.5 [189], and scikit-learn libraries [407] for model building. To evaluate the proposed prediction approach, we use 5-fold cross-validation [268], which is a validation technique to assess the predictive performance of the machine learning models and to judge how models perform to an unseen dataset [444].

During the training phase, we first discover the best value of parameters for RF, GBDT, and LSTMs using the hold-out method [262]. Then, we conduct a grid search on these values of parameters to find the best combination that achieves the highest performance. During the testing phase, the first step is to determine how long the prediction horizon should be. After conducting a series of sensitivity studies showing the changes in the value of loss function which captures the error of the model on the training dataset, we choose 15 seconds as our prediction horizon so that it is able to predict failures for the next 1500 data points. The second step is to measure the wellness of our prediction approaches. We take F-measure as the priority criterion, which is the harmonic average of precision and recall. We also consider other measure criteria simultaneously, such as precision, recall, and accuracy.

ALONE Experiment Results For each method, we conduct the testing phase five times and get the precision, recall, accuracy, and F-measure for each time. We calculate the average values of these evaluation scores of five times, and the average values are shown in Fig. 3.3.

Observations of ALONE Experiments Based on our experimental results, we have the following observations:

- Excluding driver behavior metrics results in around 8% reduction in the average F-measure (shown in Fig. 3.3).
- LSTMs outperforms RF and GBDT in both two groups.

The first observation shows that driver behavior metrics such as speed, acceleration, and steering are potentially good indicators of the failures of EV battery and associated accessories. However, driving is significantly influenced by the current driver, weather, location, and traffic conditions, which requires a personalized model to capture the real-time driving pattern.

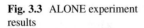

Fig. 3.3 ALONE experiment results

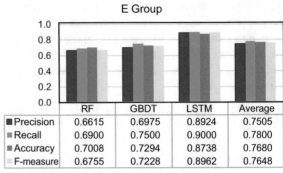

E Group

	RF	GBDT	LSTM	Average
■ Precision	0.6615	0.6975	0.8924	0.7505
■ Recall	0.6900	0.7500	0.9000	0.7800
■ Accuracy	0.7008	0.7294	0.8738	0.7680
F-measure	0.6755	0.7228	0.8962	0.7648

■ Precision   ■ Recall   ■ Accuracy   ▪ F-measure

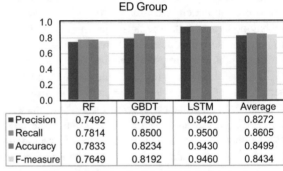

ED Group

	RF	GBDT	LSTM	Average
■ Precision	0.7492	0.7905	0.9420	0.8272
■ Recall	0.7814	0.8500	0.9500	0.8605
■ Accuracy	0.7833	0.8234	0.9430	0.8499
F-measure	0.7649	0.8192	0.9460	0.8434

■ Precision   ■ Recall   ■ Accuracy   ▪ F-measure

As to the second observation, although the degradation process of EV failures is complicated, battery and associated accessories do have their hidden failure patterns [201, 569]. As the more sophisticated neural networks including historical information, LSTMs are good at capturing hidden patterns of battery failures based on historical data and generating a sequence of predictive data points to predict the incoming patterns.

3.1.4.4   CLONE_*training* Framework

As shown in Fig. 3.4, each edge node (vehicle) is responsible for continuously performing training locally based on its private data. When a single edge node (vehicle) finishes one epoch [182], which refers to the number of iterations related to the total input dataset, it will push the value of current parameters to the Parameter EdgeServer, where the parameter values are aggregated to compute the weighted average value. Then, each edge node (vehicle) will immediately pull the updated parameter values from the Parameter EdgeServer and set the updated parameters as their current parameters to start the next epoch. The above steps will be repeated as necessary.

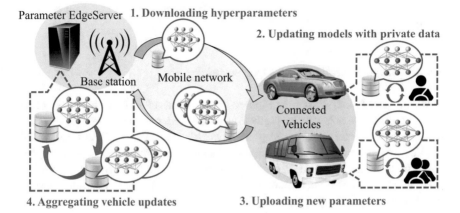

Fig. 3.4 The framework of CLONE_*training*. Each vehicle trains the neural network model locally based on its private data. Then, the value of current parameters from each vehicle is uploaded to the Parameter EdgeServer, where those parameters are aggregated and sent back to vehicles

Note that when a new edge node (vehicle) joins in, it will pull the current aggregated parameters from the Parameter EdgeServer first and set them as the initial parameters for the first round of training, which is able to speed up the training process for unseen edge nodes (vehicles). Besides, since it is asynchronous communication, for each edge node (vehicle), there is no need to stop and wait for other edge nodes to complete an epoch, which greatly reduces the latency. To illustrate the aggregation protocol of CLONE_ *training*, we need to introduce the loss function first, which is defined as follows:

$$Loss = \sum_i \left[\hat{y}_{(i)} * \lg(y_{(i)}) + (1 - \hat{y}_{(i)}) * \lg(1 - y_{(i)}) \right] \tag{3.1}$$

Here, $\hat{y}_i$ is the predicted output of a machine learning model, and the scalar y_i is the desired output of the model for each data sample i. We then define the formula to aggregate and update parameters as follows:

$$\begin{cases} P(p) \leftarrow \frac{Loss(v)}{Loss(p)+Loss(v)} P(p) + \frac{Loss(p)}{Loss(p)+Loss(v)} P(v) \\ Loss(p) \leftarrow Loss(v) \end{cases} \tag{3.2}$$

where P represents the value of a parameter, and $Loss$ stands for the value of the loss function. Besides, p refers to the Parameter EdgeServer, and v represents a specific edge node (vehicle). According to the above formula, if the model on an edge node achieves more accurate results (lower value of loss function), it will assign a higher weight to the parameters uploaded by this edge node, so that the required training time could be minimized efficiently to reach a certain accuracy level.

3.1.4.5   Implementation of CLONE_*training*

Experimental Goals We can see in Sect. 3.1.4.3 that driver behavior metrics have non-negligible impacts on the prediction of EV failures, and employing LSTMs could achieve better results compared with RF and GBDT. Therefore, in this section, we aim to deploy LSTMs-based collaborative learning approaches on edges based on EIC attributes and driver behavior metrics. The main goal is to construct personalized models by continuously tuning parameters in a collaborative fashion while protecting driver privacy and predicting failures in real-time.

Hardware Selection To build a heterogeneous hardware cluster to represent different models of EVs, we adopt two different types of hardware—Intel FRD and Jetson TX2, with different CPUs, memory, and so on. More specifically, we choose one Intel FRD as the Parameter EdgeServer, and we treat other two Intel FRDs and one Jetson TX2 as the edge nodes for vehicles to continuously "learn" latent patterns.

Experiment Setup In Sect. 3.1.4.3, we trained an accurate LSTMs model with 4 layers on the front and followed by a fully connected layer (dense layer). Now, we aim to deploy a distributed LSTMs in a collaborative fashion with the same number of layers (same hyperparameters) on edge nodes. We first distribute our whole dataset into three edge nodes, so that each edge node (vehicle) has its locally private dataset. Note that the data collection periods of three EVs are not the same, which means that the dataset is not evenly distributed.

Figure 3.5 shows the parameter distribution of the LSTMs model on the first two LSTMs layers (marked as LSTM_1 and LSTM_2) and the last fully connected layer (labeled as Dense_1). The "kernel" and "recurrent_kernel" are the parameter vectors, and the number inside angle brackets represents the shape (size) of parameters for each vector. We can see that parameter vectors are usually high-dimensional, *i.e.,* there are a huge amount of parameters. For example, $\langle 16 \times 400 \rangle$ indicates that there are 16×400 of parameters. Our whole network contains up to 297,700 parameters, including the weights and the biases. Weight is able to reflect the strength of the connection between input and output. Bias shows how

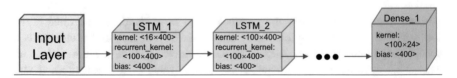

Fig. 3.5 Model parameters

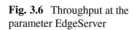

Fig. 3.6 Throughput at the parameter EdgeServer

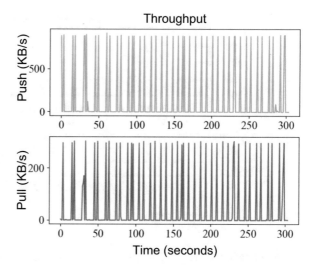

far off the predictions are from the real values. During each epoch, edge nodes (vehicles) push all parameters to the Parameter EdgeServer to get the updated value of parameters. Same as Sect. 3.1.4.3, we use a 5-fold cross-validation method to evaluate experiment results of CLONE, which will be present in Sect. 3.1.4.6 to have a direct comparison with ALONE.

Throughput   Figure 3.6 shows the throughput at the Parameter EdgeServer when three work nodes are working together. The Parameter EdgeServer receives parameters during the push process and sends parameters during the pull process. We can see that the data throughput peak is relatively stable and the peak appears intermittently. Besides, the maximum throughput for the push and pull process is around 750 KB/s and 250 KB/s, indicating that there is no big pressure on the network throughput. Figure 3.6 also proves that the push process is usually much slower than the pull process for CLONE_*training*, which was concluded by the work of [270]. This observation shows the importance to investigate methods that are able to reduce the communication latency of the push process in the future work.

3.1.4.6   Comparison Between CLONE_*training* and ALONE

In this subsection, we present the experimental results of CLONE_*training* and compare it with the algorithm performance of ALONE in two aspects: (i) training time, and (ii) evaluation scores including precision, recall, accuracy, and F-measure.

   To have a clear comparison, we conduct experiments on three experimental groups. The first group is ALONE, and we set the epoch of ALONE equal to 210. The second group is CLONE with the epoch of 70 for each edge node, and we label it as CLONE1. Since there are three edge nodes in CLONE1, the equivalent number

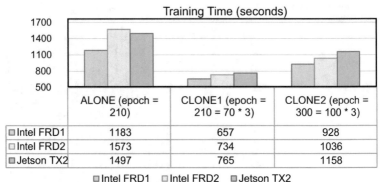

Training Time (seconds)			
	ALONE (epoch = 210)	CLONE1 (epoch = 210 = 70 * 3)	CLONE2 (epoch = 300 = 100 * 3)
☐ Intel FRD1	1183	657	928
☐ Intel FRD2	1573	734	1036
☐ Jetson TX2	1497	765	1158

☐ Intel FRD1   ☐ Intel FRD2   ☐ Jetson TX2

Fig. 3.7 Training time comparison

of iterations in total is also 210 (70 × 3). As to the third group (CLONE2), the epoch is 100 for a single edge node, which results in 300 (100 × 3) of total iterations.

Training Time Comparison We first profile and compare how the training time is spent on the three experimental groups, which is shown in Fig. 3.7.

For ALONE, the used training time varies with different edge devices—it takes 1183s and 1573s on two Intel FRDs, respectively, while taking 1497s to execute the training task on Jetson TX2. As to CLONE1, the training time of each edge node is much lower than the training time of the single edge node of ALONE. Since there are three edge nodes in CLONE1, the training time of CLONE1 should be one-third of ALONE theoretically. However, due to the inevitable delay of the parameter transmissions, the training time of CLONE1 is greater than one-third of ALONE. We then increase the epoch value from 70 × 3 to 100 × 3 (CLONE2), and it can be seen that the required training time is longer than CLONE1 as it has a larger number of iterations related to the input dataset during the training phase, but it still less than ALONE which has a lower epoch value. *Note that with the participation of more edge nodes and larger size of the input dataset, the advantages of CLONE_training in training time reduction will be more obvious.* Note that the training time in Intel FRD1 and FRD2 are different, which might be due to the difference in the time length of usage, system environment, and the distance to the router.

Evaluation Score Comparison We then calculate the average evaluation scores for each group, which is shown in Fig. 3.8. Comparing ALONE and CLONE1, we can see that the overall evaluation scores of CLONE1 are lower than ALONE. This could be caused by the fact that the prediction accuracy will be improved with the increasing number of iterations passing the full dataset through the current model, and ALONE has a higher epoch value (210) than CLONE1 (70). However, the evaluation score of CLONE could be further improved by increasing the epoch value, *e.g.,* the evaluation scores of ALONE and CLONE2 are relatively equal.

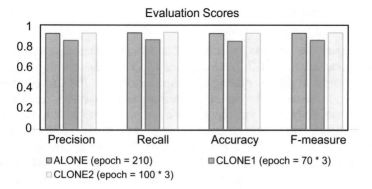

Fig. 3.8 Algorithm performance

Besides, in the CLONE setting, due to the hardware difference, powerful edge nodes that have a higher FLOPS may train the model faster than other edge nodes. Here, FLOPS is an abbreviation for floating point operations per second ("S" stands for the time unit, *i.e.,* second), which is often used to estimate the computation power of a device (greater GFLOPS usually indicates a more powerful computation resource and faster computation speed). Therefore, during the same training time period, powerful edge nodes can achieve a higher value of epoch, *i.e.,* finish more training passes than other edge nodes; therefore, the output of a powerful edge node will make a greater contribution to the training of the global model, which leads to better performance when the trained model performs inference on the data generated by the powerful node. When the parameters of the poor training results are uploaded to the Parameter EdgeServer, the global accuracy of CLONE1 will be influenced. This may explain the performance gap between ALONE and CLONE1 whose total epoch values are the same.

However, when we further increase the value of epoch (CLONE2), it is capable to achieve high evaluation scores as ALONE. Note that, by observing Fig. 3.7, the training time of CLONE2 is much lower than ALONE, even though CLONE2 has a higher epoch.

3.1.4.7   Discussion

Compared with ALONE, CLONE_*training* is able to reduce model training time without sacrificing algorithm performance. With more edge nodes involved, the advantages of CLONE in training time reduction will be more obvious. Besides, CLONE_*training* provides personalized models to predict the failures of EV battery and associated accessories considering the current driver behaviors, and CLONE_*training* is capable to speed up the analysis tasks while protecting user privacy better as it does not need to transfer any portion of the sensitive dataset via the network.

3.1.5   Summary

In this chapter, we proposed a collaborative learning framework on the edges (CLONE), including CLONE_ *training* and CLONE_*inference*. It demonstrated the effectiveness of privacy serving and latency reduction. For CLONE_*training*, we chose the failure prediction of EV battery and associated components as our first case study, and the corresponding experimental results showed that CLONE_*training* could reduce model training time without sacrificing algorithm performance. Besides, we found that adding driver behavior metrics could improve the prediction accuracy for the EV failure prediction. As to CLONE_*inference*, customer tracking in a grocery store was selected as the second case study. We presented a detailed description of how the CLONE_*inference* solution could be employed to solve the multi-target multi-camera tracking problem.

There are some possible improvements for CLONE. We list three of them for the discussion. (*i*) *Bandwidth demand*: as the increasing number of edge nodes or the participation of larger neural networks, the communication of CLONE may be limited by bandwidth. In this context, we could leverage the Parameter EdgeServer group. In the group, Parameter EdgeServers communicate with each other. Each Parameter EdgeServer is only responsible for a portion of parameters, and they work together to maintain globally shared parameters and their updates. (*ii*) *Aggregation protocol:* it is essential to find a suitable aggregation rule for the Parameter EdgeServer to aggregate parameters, which requires excessive experiments based on the specific experimental conditions. (*iii*) *Push/pull latency:* as to CLONE_*training*, pushing parameters to the Parameter EdgeServer is usually much slower than pulling parameters. As to CLONE_*inference*, the case is different. It is essential to investigate methods to reduce push/pull latency (possible solutions include structured updates and sketched updates [270]).

Failure prediction of EV battery is used as a case study to show how CLONE solution could be employed in the training stage and the inference stage. There are a variety of other meaningful use cases that CLONE could help, particularly for two types of scenarios: (*i*) real-time applications that require developing suitable machine learning algorithms on the edges, and (*ii*) due to the privacy or/and the large network bandwidth constraints, the training dataset cannot be moved away from its source.

3.2   Collaborative Cloud-Edge Computation for Personalized Driving Behavior Modeling

3.2.1   Introduction

The past decades have seen significant improvements in road safety. However, numerous traffic accidents are still occurring every day. It is estimated that over

1.2 million people worldwide die in road crashes each year, with millions more sustaining serious injuries and living with long-term adverse health consequences [396]. Most traffic accidents are caused by human errors, such as speeding, distracted driving, drowsy driving, and drunk driving.

Advanced driver assistance system (ADAS) is developed to automate, adapt, and enhance vehicle systems to increase road safety. ADAS provides various safety features including blind spot monitoring, lane change assistance, and forward collision warnings. Driving behavior modeling is an essential component in ADAS to detect abnormal driving behaviors, send early alerts, and therefore reduce accidents [600]. Driving behavior modeling can also be used by insurance companies to determine the vehicle insurance premium [58, 373].

Many methods have been proposed to model driving behaviors and detect anomalies. For example, the driver's facial features, such as the state of eyes and mouth, are used in classification methods to detect whether the driver is drowsy or distracted [246]. These methods usually require drivers to wear glasses or leverage in-car cameras. Driving data, such as vehicle velocity, angular velocity, and acceleration, is also used in driving behavior analysis [333, 585]. These data are often collected through on-board diagnostic (OBD) sensors or smart phones. While many research efforts have been devoted to driving behavior analysis, several open challenges still remain when deploying the driving behavior models in real practice:

(1) Personalization Existing driving behavior models are usually trained on generic datasets, which do not consider driving contextual information, such as drivers' individual difference, location, weather, and traffic conditions. However, driving is significantly influenced by these factors, which lead to personalized driving behaviors. Drivers may have distinct driving behaviors because of their individual difference, such as age group, gender, and driving experience. For example, the driving behavior model that is built on datasets from experienced drivers cannot appropriately capture the behavior of novice drivers. The same driver may have different behaviors under various situations. For example, the driving speed of 60 miles per hour on US highways is normal in sunny days. However, it is dangerous in blizzard. If the driving behavior model is only trained on data under normal weather conditions, it will not be able to detect abnormal behaviors under severe weather conditions.

(2) Latency Real-time performance is a stringent requirement for ADAS. For example, fatigue driving or other abnormal driving behaviors should be detected immediately, in order to avoid accidents or minimize damage. There are two types of methods to run the driving behavior model: cloud-based and edge-based methods. For the cloud-based method, the computing unit sends the driving behavior data to the model in the cloud and then receives decisions over the network [588]. This method suffers greatly from the stability and latency of the network. For the edge-based method, ADAS analyzes the driving behavior on board, which can minimize the network latency. However, the computing unit on vehicle usually has limited computation and storage resources, which are not sufficient to run a large driving behavior model.

(3) Privacy As General Data Protection Regulation (GDPR) [437] has already taken effect, the collection of driving data as well as the modeling of driving behavior need to consider various privacy regulations. However, most existing approaches usually collect users' driving data, send them directly to the cloud, build the model and then conduct analysis. This process may put the user privacy at risk.

(4) Integration The driving data is being collected by various applications for their own driving behavior analysis. For example, the insurance company analyzes driving data to dynamically adjust the insured's premium. The ride sharing company detects abnormal driving behaviors, in order to increase ride safety. It is desired to have an integrated driving behavior model that can provide a universal service to various applications. The efforts on data collection and computational analysis can therefore be significantly reduced.

To address these challenges, we propose pBEAM, a collaborative cloud-edge computation system for personalized driving behavior modeling. Comparing with the traditional cloud-based approaches which upload all the data to a centralized cloud service, edge computing can reduce or minimize network latency [461, 483] and address privacy concerns [343]. Vehicles are important players in edge computing, which are suitable to perform certain computational tasks [549] when building driving behavior models. Figure 3.9 shows the high level work flow of our method. Specifically, we first train a common baseline model in the cloud using all the

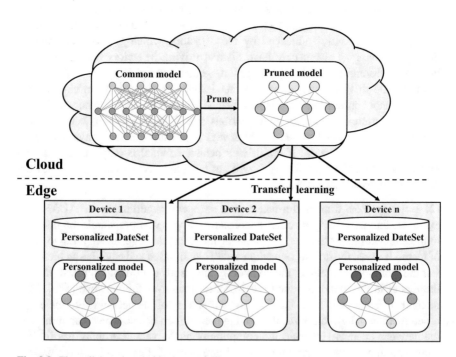

Fig. 3.9 The collaborative cloud-edge method

driving data, such as velocity, orientation, and acceleration over time. There data are anonymized and integrated from users who would like to share their data in the cloud. The cloud model is based on Generative Adversarial Recurrent Neural Networks (GARNN), which adapts to the dynamic change of users' normal driving. This deep neural network model is further pruned or compressed to a smaller model and transferred to the edge device on each vehicle. After compressing or pruning, the computational cost of running the deep learning model can be reduced, thereby satisfying the computing power restrictions of the on-board hardware. We then train a personalized model on top of the pruned model through transfer learning, considering the specific driving condition or context information.

There are several advantages of our method. First, the Generative Adversarial Recurrent Neural Networks (GARNN) are adaptive to the dynamic change of normal driving. Instead of only training a binary classifier to identify abnormal driving behaviors, GARNN learns both a generator that tries to produce true driving data as well as a discriminator that tries to distinguish between generated data and true driving data. Essentially, GARNN learns what the true driving data should look like. Any abnormal driving behaviors that do not follow the distribution of true driving data will be considered as anomalies. The labeling of anomalous data is not needed when training the model, and is only used for the model performance evaluation. Second, the transfer learning from cloud to edge improves the model performance and robustness. The pruning of deep neural networks minimizes the transferring load of the model while maximally preserves its original performance. Third, the model trained on edge devices outperforms the original cloud model, since personal and contextual information are considered in the training. Last but not the least, users' privacy is well protected because they do not need to upload personal data to the cloud.

We summarize the contributions of this work as follows:

- A personal driving behavior modeling system, pBEAM, is presented. pBEAM can improve the overall performance of detecting abnormal driving behaviors and reduce detection latency, while ensure the privacy of drivers. A RESTful engine in pBEAM provides universal modeling services to different applications.
- A deep learning model, CGARNN-Edge, is developed to model the driving behavior. Experimental results on driving data from both real world and a driving simulator show that the proposed CGARNN-Edge achieves the best performance among all the methods.
- A collaborative cloud-edge computation method that trains and prunes common models in the cloud and conducts transferring learning on the edge is developed. This method is general and can be used in other cloud-edge applications.

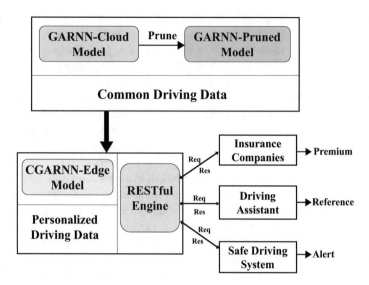

Fig. 3.10 The overview of pBEAM. pBEAM includes four parts: GARNN-Cloud, GARNN-Pruned, CGARNN-Edge, and RESTful Engine

3.2.2   System Design

In this section, we introduce pBEAM, a collaborative cloud-edge computation system for personalized driving behavior modeling. The traditional cloud-based methods require the uploading of driving data to the cloud, which may cause privacy issues as the driving data usually contain personal information. In addition, it is not efficient or even feasible to store all the data and perform model training on edge devices as the storage and computing resources are limited.

To address these challenges, we propose pBEAM in this chapter. As shown in Fig. 3.10, it has four stages to build pBEAM: (1) build a common baseline model in the cloud using all the driving data, denoted by GARNN-Cloud, (2) prune the baseline cloud model to reduce the total number of parameters and model size, denoted by GARNN-Pruned, (3) transfer the pruned model to the edge device and retrain an edge model by considering conditional or contextual information, denoted by CGARNN-Edge, and (4) provide RESTful web services for third-party applications development, denoted by RESTful Engine. RESTful Engine is designed on top of CGARNN-Edge and provides normalized APIs to developers. For example, the CGARNN-Edge model can be used in the driving assistance systems to detect abnormal driving behaviors. In practice, GARNN-Cloud and GARNN-Pruned are located in the cloud servers in order to leverage the large-scale storage and computation infrastructure. The pruning of GARNN-Cloud helps to reduce the cost of training CGARNN-Edge on the edge. CGARNN-Edge and RESTful engine are deployed on the vehicle computing unit to train on personalized data and reduce the amount of data to be uploaded to the cloud. We will present the

Fig. 3.11 The architecture of GARNN-Cloud

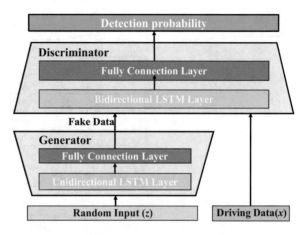

details of GARNN-Cloud, GARNN-Pruned, CGARNN-Edge, and RESTful Engine in the following sections.

3.2.3 GARNN-Cloud

To model the common driving patterns, we first build a baseline cloud model using Generative Adversarial Recurrent Neural Networks (GARNN) which is inspired by C-RNN-GAN [359]. GARNN is a continuous recurrent neural network with adversarial training. The overall architecture is shown in Fig. 3.11. GARNN consists of two components: generator (G) and discriminator (D), which are built based on LSTM (Long Short-Term Memory networks), a type of recurrent neural networks [214]. The generator is trained to generate data that is indistinguishable from real normal driving data, while the discriminator is trained to identify whether the generated data is real or not. Note that since the driving data is time series, we use unidirectional LSTM in the generator to capture the temporal direction. In the discriminator, we instead use a bidirectional LSTM as the goal is to classify the driving data without the constrain of a particular sequential order.

The input of discriminator includes the fake data generated by generator and the real driving data. The loss functions of generator and discriminator, L_D and L_G, are defined as follows:

$$L_G = \frac{1}{m} \sum_{i=1}^{m} \log(1 - D(G(z^i)))\,,$$

$$L_D = \frac{1}{m} \sum_{i=1}^{m} \left[-\log D(x^i) - (\log(1 - D(G(z^i)))) \right]\,,$$

Fig. 3.12 The process of model pruning

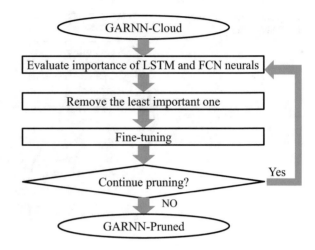

where $z^{(i)}$ is a time series of uniform random vectors in $[0, 1]^k$, k is the feature size, and $x^{(i)}$ is the time series of true driving data, normalized between 0 and 1.

3.2.4 GARNN-Pruned

When the cloud model is finished training, we use pruning to reduce the total model size so that the model can be transferred to the edge device which has limited storage and computing resources. As shown in Fig. 3.12, we prune the connections with low-absolute values. All connections with weights below a threshold are removed from the network, therefore converting a dense network into a sparse one. The threshold is a hyper-parameter that depends on the trade-off curve between compression ratio and prediction accuracy. We use the mask mechanism to implement model pruning; weights below the threshold are masked by zeros, while those above are masked by ones. By taking the dot product between the original weight tensor and the mask tensor, the connections with weights smaller than the threshold are set to zero, i.e., pruned.

We apply an automated gradual pruning algorithm [613] to cut the unimportant connections. Over a span of n pruning steps, the sparsity is increased from an initial sparsity value s_i to a final sparsity value s_f. The sparsity at time t is defined as follows:

$$s_t = s_f + \left(s_i - s_f\right)\left(1 - \frac{t - t_0}{n\Delta t}\right)^3$$

where t_0 is the start step and Δt is the pruning frequency. In this chapter, we start pruning the model at the first step and empirically set the pruning frequency to be 10. The initial sparsity s_i is 0 and the target sparsity s_f is set to 0.5. As the sparsity of the network increases, the weight masks are updated. Once the model's sparsity s_t achieves the target sparsity, the weight masks are no longer updated and the pruning process is completed. Comparing with the original cloud model, the pruned model may suffer an accuracy loss, but achieve a considerable reduction in the model size.

3.2.5   CGARNN-Edge

After the pruned model is transferred to the edge device, we will train a personalized model on top of GARNN-Pruned through transfer learning, considering the contextual information. Specifically, GARNN-Pruned is used as the initialization of the personalized model. We propose to use CGARNN as the personalized model, where both the generator and discriminator are conditioned on the context information y. y can be any kind of contextual information, such as drivers' personal data, location, weather condition, and traffic situation. We perform the conditioning by feeding y into both discriminator and generator as additional input. Figure 3.13 shows the architecture of CGARNN-Edge.

The new loss functions of generator and discriminator, L_D and L_G, are defined as follows:

$$L_G = \frac{1}{m} \sum_{i=1}^{m} \log(1 - D(G(z^i|y)))\,,$$

$$L_D = \frac{1}{m} \sum_{i=1}^{m} \left[-\log D(x^i|y) - (\log(1 - D(G(z^i|y)))) \right]$$

where the random input noise z and y are combined as a joint hidden representation in the generator, while x and y are combined as new input to the discriminator.

As shown in Fig. 3.13, CGARNN-Edge will be deployed on the computing platform of the vehicle and trained on the personalized data. After the model training, the discriminator of CGARNN-Edge can be used to detect whether the driving behavior is normal or not. The input is the real-time driving behavior and the output is the probability of being an anomaly.

Recent research work is devoted to train the machine learning models on the edge based on the local data, along the same line as CGARNN-Edge [91]. However, one of the most significant challenges of training model on the edge is training data labeling [138]. For example, it is difficult to train an object detection model on the vehicle with images generated from the on-board camera, which requires image labelling in real-time. CGARNN-Edge does not need the labeled data at the training stage as it only uses the driver's normal driving behavior as the input. The drivers' personal data (such as age group and gender) is the one-time input and CGARNN-Edge can obtain it at the beginning. CGARNN-Edge can acquire the information of

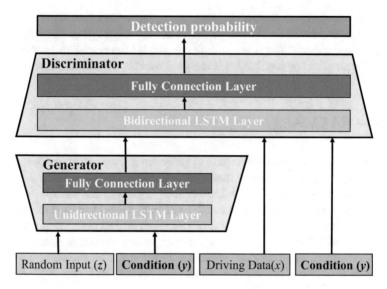

Fig. 3.13 The architecture of CGARNN-Edge

location, weather condition, and traffic situation by using web services, such as the weather report services and the Google maps API.

3.2.6 RESTful Engine

After CGARNN-Edge is trained, the RESTful Engine will call for it and provide web service for the third-party applications. Figure 3.10 lists three potential services: (1) insurance companies leverage the driving behavior analysis for credit rating and premium adjustment, (2) driving assistant provides driving reference, and (3) ridesharing companies can detect abnormal driving behaviors to increase ride safety [313] and send alarms.

Figure 3.14 lists two RESTful APIs and their purposes when using GET operations. CGARNN-Edge is represented by a URI which consists of four fields. The first field is the IP address and port number of the vehicle. The second field represents the particular driving behavior model, i.e., CGARNN-Edge. The third field indicates the request type, such as real-time and batch. The fourth field is the arguments which will be sent to the RESTful Engine. For example, if the driving assistant system needs to obtain the driving behavior in the real-time manner for the sake of safety, it should follow the URI shown in *Example 1* in Fig. 3.14 and send the timestamp argument. The response is the detection probability that the driving behavior is normal or not at that moment. If an insurance company wants to know the driving performance of a customer in the last month, it needs to follow the URI

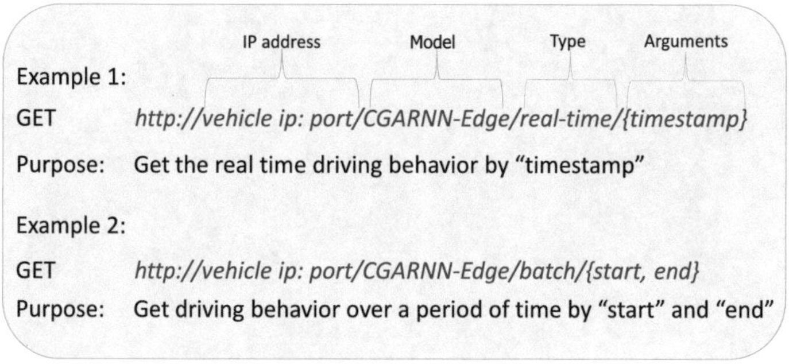

Fig. 3.14 The examples of RESTful Engine

shown in *Example 2* and sends the starting and ending timestamp. The response is the probability of the driving behavior being abnormal during that period.

The design of RESTful Engine can address the challenge of integration. In this case, there is no need to build a driving behavior model for each application separately. The integration function of the RESTful Engine and the privacy-preserving goal of the proposed system do not conflict. The original driving data and the trained CGARNN-Edge will be kept on the vehicle and not be transferred. The third-party application can only call the model via the RESTful API. Moreover, the access permission of the RESTful Engine is managed by the driver, which will protect user privacy.

3.2.7   Summary

This Section presents a collaborative cloud-edge computation method for personalized driving behavior modeling. This method models driving behaviors using Generative Adversarial Recurrent Neural Networks, which is adaptive to the dynamic change of normal driving. To address the challenges of personalization and privacy, the proposed collaborative cloud-edge computation method first trains a common baseline model in the cloud and then trains a personalized model on the edge through transfer learning, considering personal and contextual information as additional conditions. Model pruning is applied to minimize the transferring load as well as to maximally preserve the original cloud model performance. The proposed CGARNN-Edge model achieves the best performance among all the models.

Chapter 4
Systems Runtime Optimization

4.1 E2M: An Energy-Efficient Middleware for Autonomous Mobile Robots

4.1.1 Introduction

An autonomous mobile robot (AMR) is mainly composed of mechanical parts (e.g., wheels, engine, etc.), sensors (e.g., camera), and computing hardware (e.g., CPU, GPU) that allow the robot to autonomously drive and perform a predefined set of jobs. Due to their relatively low cost, high flexibility, and high reliability, various AMRs have been designed and employed in several industry applications [28, 123, 288]. For example, an AMR can be programmed to patrol the fences of a private area so that the security team can be timely notified of intrusions [314]. Typical jobs for AMRs include security patrol, object detection, and face recognition [488]. However, the correct operation of an AMR is strictly dependent on its limited battery life. Therefore, it is important to ensure high energy efficiency to maximize the battery life of AMRs.

Some previous studies have focused on reducing the energy consumption due to the mechanical parts of the AMR. For example, they propose solutions to dynamically find the most energy-efficient path to reach a certain location [60, 115, 203, 352]. However, these solutions do not take into consideration the computational portion of the AMR's energy consumption. In fact, the computing resources of a typical AMR can account for the 33% of the total energy consumption. The most common applications of AMRs use machine learning models in computer vision applications. Thus, in order to improve the energy efficiency of the computing resources of an AMR, it is important to focus on the execution of computer vision-based applications. Unfortunately, most previous research studies on such applications either focus mainly on performance [292, 305, 531] or employ pruning and compression techniques to make the trained model more energy efficient [577].

© The Author(s), under exclusive license to Springer Nature Switzerland AG 2021
W. Shi, L. Liu, *Computing Systems for Autonomous Driving*,
https://doi.org/10.1007/978-3-030-81564-6_4

To the best of our knowledge, no previous work has yet studied the energy efficiency of AMRs during the execution of computer vision-based applications.

Due to the above described shortcomings of the existing literature on AMRs, in this chapter we first conduct an in-depth study of the computer vision application execution on AMRs to profile its energy consumption and to discover the main sources of inefficiency. We find two main sources of high energy consumption across various applications: access to sensor data and model inference. Accordingly, we find **three main inefficiencies** for these two energy sources. The first is **uncoordinated access to sensor data.** Each computer vision process directly interacts with the sensors, e.g., camera, of the AMR to acquire data, e.g., camera frames. As a result, N concurrently executing processes may acquire N camera frames for inference within a short period of time. If the N acquired frames are similar to one another, using one of the N frames for all the concurrent processes would not change their inference results, which leads to energy waste for acquiring more camera frames than necessary. The second one is **performance-oriented model inference execution.** A computer vision process on an AMR continuously performs frame acquisition and model inference without waiting time, i.e., without delaying the next frame acquisition and inference time. Although this execution mode ensures high performance (e.g., safety, accuracy), it also prevents the computing hardware (e.g., CPU, GPU) to reach deep-sleep states for energy savings. And the third is **uncoordinated execution of concurrent jobs.** Multiple executing inference processes executing concurrently, even with a per-process optimized waiting time, may still prevent the computing hardware from reaching the deep-sleep state. In fact, most computing systems can reach the deep-sleep state only after a certain amount of time has passed. Thus, uncoordinated waiting times may cause the computing hardware to never reach the necessary contiguous idle time to enter the sleep state.

In order to address the above described inefficiencies, we propose E2M, a generalized energy-efficient middleware for autonomous mobile robots. E2M consists of four major components: sensor buffer, performance analyzer, energy saver, and coordinator. The sensor buffer is designed to capture the sensor data (e.g., camera frames). It allows to coordinate the concurrent access to sensor data and reduce the total amount of data collected for energy savings. The energy saver profiles the energy consumption of the executing processes based on their waiting times. The performance analyzer profiles the performance of each running process based on a predefined per-process metric and desired target. The coordinator collects the information about energy consumption and performance analysis to find the best waiting times for each process that maximize the contiguous idle time of the computing hardware for energy savings while ensuring good performance for each process. In practice, the coordinator controls the waiting time by controlling the feed time of sensor data to each process.

In summary, this work makes the following three contributions:

- We analyze the computational energy consumption of autonomous mobile robots and find three main sources of inefficiency: uncoordinated access to sensor data, performance-oriented model inference execution, and uncoordinated execution of concurrent jobs.

- We propose an energy-efficient middleware (E2M) for autonomous mobile robots to fix the three inefficiencies. E2M coordinates the access of processes to sensor data and also coordinates the execution of the processes to maximize energy savings while ensuring good performance.
- We develop a prototype of E2M on a real-world AMR and test it on a real scenario. Our experimental results show that E2M leads to 24% energy savings for the computing platform, which translates into an extra 11.5% of battery time and 14 extra minutes of robot runtime, with a performance degradation lower than 7.9% for safety and 1.84% for accuracy.

4.1.2   Motivation

The deep learning-based approach has been widely used in autonomous driving applications. However, currently most of the proposed methods are performance driven without considering the energy consumption. In this section, we first describe our experimental setup used to execute the experiments. Second, we analyze the power breakdown of a real AMR platform. Third, we analyze the power breakdown of computer vision processes to identify the highest sources of energy consumption. Finally, we analyze the effect of the waiting time of processes on the computing power dissipation.

4.1.2.1   Experimental Setup

In order to characterize the energy consumption of autonomous mobile robots, we analyze an indoor autonomous mobile robot called HydraOne [548]. The HydraOne platform is shown in Fig. 4.1b. Different from many heavy-weight AMRs used for

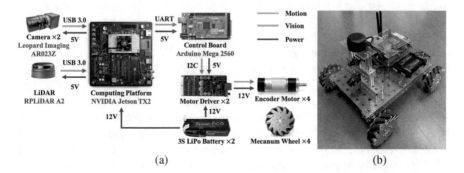

(a)   (b)

Fig. 4.1 (**a**) The hardware design of HydraOne platform. (**b**) The HydraOne platform

one specific application (e.g., moving heavy objects), we consider multi-purpose AMRs, which can run various computer vision applications concurrently. The user can decide which applications to execute at any time. For example, multi-purpose AMRs can be used in retail stores to help managers execute applications such as understanding out-of-stock items, guaranteeing price integrity, confirming product showcases, and identifying hazardous conditions [514]. Thus, multi-purpose AMRs do not necessarily need to be heavy weight. The HydraOne configuration is a representative design of such robots with multiple concurrent computer vision applications running on it. In particular, HydraOne is a full-stack research and education platform and includes mechanical components, vision sensors, computing hardware, and communication system. All resources on HydraOne are managed by the Robot Operating System (ROS). Figure 4.1a shows the hardware design of HydraOne. Two leopard cameras [289] and an RPLiDAR [487] are connected to the computing platform via USB cable. RPLiDAR is a 2D laser scanner that provides 360° laser range scanning. The results of a set of data points in space are called *point cloud*. An Nvidia Jetson TX2 board is used as the computing platform [391]. One Arduino Mega 2560 board with two motor driver boards is used to control HydraOne. Two 3S Lipo batteries are used to power the whole system: one for the computing platform and the other for the wheels. The capacity of each 3S Lipo battery is 5000 mAh.

On top of HydraOne, we implement a deep learning-based end-to-end free space detection application and an object detection application. Unfortunately, existing navigation applications cannot be used in our experiments because they are not designed and trained in our environment. As a result, we designed and trained a convolutional neural network (CNN) called HydraNet to achieve end-to-end free space detection, where the input is the frame from camera and the output is the controlling command, i.e., linear and angular speed to the robot. The training dataset, which contains the image frames labeled with control messages, is obtained by a human remotely controlling HydraOne. There are five convolutional layers and four fully connected network (FCN) layers in HydraNet. The convolutional layers are designed to perform feature extraction. The first three convolutional layers have a 5×5 kernel and a 2×2 stride. The last two convolutional layers have a non-strided convolution with a 3×3 kernel. The filters of these five layers are 24, 36, 48, 64, and 64, respectively. Four FCN layers are designed as the decision maker for the driving and lead to two output values of linear and angular speed. The number of cells in each FCN layer is 512, 100, 50, and 10, respectively. Based on the experience of industry practice and a series of experiments, we choose Rectified Linear Unit (ReLU) as the activation function for all nine layers. The RPLidar is used in this context to collect data about object distances. For object detection, we choose the combination of MobileNet and Single Shot MultiBox Detector (SSD) [320]. Through replacing the lightweight depthwise separable convolution layer with standard convolution layer to reduce the number of computations, MobileNet becomes more suitable for the resource-constrained AMR platform [222]. Furthermore, SSD is a widely used deep learning model for objection detection.

The power dissipation of the computing platform and sensors is measured using Watts Up Pro Electricity Consumption Meter [210], which records both the current and the real power every second. The power dissipation of the wheels is measured using the Lipo battery charger, which records the energy consumption. The error of power dissipation is less than 1% and it is ignorable. The reason why we use two different meters is that we have two batteries installed on HydraOne, one powering the locomotive mechanisms (e.g., wheels) and the other one powering sensors and computing resources. We run HydraOne with free space detection and object detection for 10 min and use the charger and the Watts Up Pro to measure the energy consumption. Then we calculate the average power dissipation based on the energy consumption and running time. When the batteries are fully charged, HydraOne platforms can run with HydraNet and MobileNet-SSD for approximately 2 h.

4.1.2.2 AMR Power Breakdown

Figure 4.2 shows the power dissipation of the entire HydraOne platform. We run HydraNet and MobileNet-SSD on HydraOne for 10 min. The total power dissipation of HydraOne is 39.1 W. From Fig. 4.2, we can see that the locomotion of HydraOne (*wheels* in the figure) consumes over half of the total power dissipation; the power dissipation of computation is 33% and that of the sensors is 11%. Out of the 33% of computational power dissipation, the model inference of HydraNet and MobileNet-SSD consumes 10% and 12% of the total power dissipation of the robot, respectively. The remaining 10% indicated with *other* in the figure includes the power dissipation of the operating system and sensor drivers.

As we can observe from this analysis, each implemented application may further increase the total computational power dissipation. Actually, in many cases,

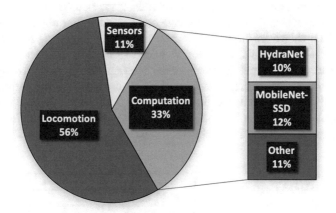

Fig. 4.2 Power dissipation breakdown of an AMR

AMRs may have to run more than two applications. For example, a surveillance AMR may implement, other than the free space detection and object recognition application, additional applications for face recognition (to timely detect the identity of intruders), self-diagnostic, and other third-party applications deemed necessary by the user. This fact leads to the following observation:

Observation 4.1 *The AMR's computational power dissipation can highly influence its autonomy. It is thus necessary to optimize the computing system of AMRs for high energy efficiency.*

4.1.2.3   Computer Vision Power Breakdown

We now analyze how the power is used by each application implemented on HydraOne. These applications use deep learning to detect free space and objects. In general, each application executes a process for the detection that works in three steps. In the first step, the process captures a frame from the camera. In the second step, the process executes model inference. In the last step, the process publishes/shows the results. The publisher/subscriber mechanism is used for data communication in ROS. We believe that these three steps are general enough to characterize most of the other applications and processes not tested in this chapter. In order to understand the energy consumption of each running process, we measure the power dissipation during each one of the above described three steps for HydraNet and MobileNet-SSD. When no waiting time of the application is applied, the power breakdown results are shown in Table 4.1.

From the power breakdown analysis in Table 4.1, we can see that capturing frames from camera consumes 2.2 W for both HydraNet and MobileNet-SSD. The power cost of model loading and inference is 0.8 W for HydraNet and 2.5 W for MobileNet-SSD. The reason why MobileNet-SSD consumes more energy than HydraNet is that MobileNet-SSD has more layers and neural cells than HydraNet. After model inference, the results of HydraNet are published to another ROS node, which executes the control message on HydraOne. For object detection, the result will be shown to the user. The power cost for publishing/showing results is 0.9 W for HydraNet and 0.1 W for MobileNet-SSD. The difference is owing to the overhead of ROS's publisher/subscriber mechanism. Thus, on average across different processes, capturing frames costs 2.2 W, running the model inference costs 1.65 W, and publishing/showing the results costs 0.5 W. An important observation must be made about the application access to camera frames. Currently, applications access the camera directly and exclusively, which means that when an application is

Table 4.1 Power dissipation breakdown of HydraNet and MobileNet-SSD

Power dissipation (W)	HydraNet	MobileNet-SSD
Capturing frame	2.2	2.2
Model inference	0.8	2.5
Publish results	0.9	0.1

accessing the camera, other applications can either wait to access it or access another camera (if available) to avoid performance degradation. Thus, the total energy consumption of two applications accessing the camera frames could potentially be halved by accessing the camera once and sharing the frame among the two requesting applications. Note that the energy savings with this method increase with the number of concurrent applications. As it can be observed in Table 4.1, the frame capturing and model inference consume the 88% of the total process power dissipation. This leads to the following observation:

Observation 4.2 *High power dissipation in AMR processes is mainly due to capturing camera frames and executing the model inference. Thus, to improve the energy efficiency of AMRs, we need to focus on these two steps like coordinating access to sensor data.*

4.1.2.4 Waiting Time vs. Power Dissipation

Given the above observations, we need to find a way to improve the energy efficiency of capturing camera frames and executing the model inference for AMR processes. To do so, we explore the possibility to introduce a *waiting time* between two consecutive inferences of computer vision processes. The waiting time is defined as the amount of time between the end of a model inference and the start of the next one. Currently these processes execute inferences continuously without delaying the next frame acquisition and inference time. Although this execution mode ensures high performance (e.g., safety, accuracy), it also prevents the computing hardware (e.g., CPU, GPU) to reach deep-sleep states for energy savings. Hence, the performance-oriented model inference is another source of inefficiency. By introducing a waiting time, we are delaying the execution of the next inference to trade off performance for energy savings. Note that the waiting time, other than reducing the energy consumption due to the inference, reduces also the number of camera frames captured. Thus, by manipulating the waiting time, we can improve the energy efficiency of the two steps in computer vision processes that consume the highest amount of energy. Here, we use the term performance to indicate in general a specific metric associated with each process. For example, the performance metric for the free space detection application is safety, i.e., distance of the robot from obstacles, while the performance metric for the object detection is accuracy, i.e., how many objects are actually recognized. It is of primary importance to meet a desired performance target while reducing the energy consumption.

To test the effect of the waiting time on energy consumption, we run HydraNet model inference and MobileNet-SSD model inference with different waiting time on the HydraOne platform and measure the power dissipation due to the computing system. We conduct the experiments with waiting time from 0 to 0.5 s. The results are shown in Fig. 4.3a and b. Introducing a waiting time in HydraNet could degrade the safety of HydraOne, i.e., the average distance of HydraOne from the surrounding objects decreases with an increasing waiting time. In relation with Fig. 4.3a, c shows

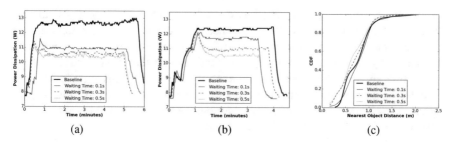

Fig. 4.3 Effect of waiting time on (**a**) HydraNet and (**b**) MobileNet-SSD. (**c**) The distance to the nearest object CDF

the distance of HydraOne from the surrounding objects for different waiting times. As the figures show, the waiting time can decrease the power dissipation of the computing platform by up to 40% for HydraNet and 35% for MobileNet-SSD with less than 0.15 m increase of distance to objects. The power consumption of HydraOne is 7.5 W when no inference is executed. Adding the HydraNet inference with zero waiting time, i.e., as the *baseline* of reference, increases the power dissipation of the platform to 12.7 W on average, which corresponds to a 41% power increase (similar for MobilNet-SSD). Introducing the waiting time can help reduce the power dissipation due to the inference. In particular, by introducing 0.1 s waiting time in HydraNet (i.e., capture frame and run inference every 100 ms), the power dissipation decreases to 11.4 W, which is 10% lower than no wait. On the other hand, further increasing the waiting time leads to smaller increases in power savings and reaches an average of 10.5 W for 0.5 s waiting time, which corresponds to a 17% power reduction. The reason for this marginal increases in power savings for an increased waiting time is that the average power consumption of the computing platform when waiting time is applied saturates to a minimum power consumption for longer waiting times. In addition, the performance degradation is acceptable when introducing small waiting times. From Fig. 4.3c, we can observe that the performance of 0.1 s waiting time is similar to that of the baseline. Even for 0.5 s waiting time, the increase of distance to objects is less than 0.15 m, which is still acceptable. Thus, high power reductions and small performance degradation can be obtained using small waiting times. Note that MobileNet-SSD shows similar behavior but has less sensitivity to increasing waiting time because MobileNet-SSD has a much longer model inference time than HydraNet. These experimental results lead us to the following observation:

Observation 4.3 *The relation between the waiting time and the power dissipation is nonlinear and shows high power reductions for small waiting times. Thus, it is possible to trade off a small performance degradation for high energy savings.*

4.1.3   Energy Efficient Middleware

The goal of the Energy Efficient Middleware (E2M) is to provide a general software solution to make the computer vision-based applications on autonomous mobile robots more energy efficient. Although there are several different computer vision applications, for simplicity here we describe the design of E2M based on two specific applications that we have fully implemented on our testbed platform HydraOne, i.e., object detection, which recognizes objects from camera frames, and end-to-end free space detection, which directly commands the speed and direction of the robot based on camera frame analysis. However, the design of E2M is general enough so that it can be easily used for any type of computer vision applications. The design presented in the next sections is based on introducing waiting time and optimizing the system overhead for frame capturing and model loading. First, we present the overview of E2M. Then, we discuss the details of each application within the E2M system.

Figure 4.4 shows the overview of E2M, which consists of four components: the sensor buffer, the performance analyzer, the energy saver, and the coordinator. E2M locates in the middle between the lower level hardware (i.e., sensors and wheels) and the applications (e.g., free space detection and object detection).

The sensor buffer communicates with the sensors including camera and RPLi-DAR to get image frames and point cloud data. Also, the linear and angular speed can be read from the wheels' ROS node. A ROS node is an executable process that uses ROS to communicate with other nodes [443]. Wheel's ROS node is used to send control messages to the motor's driver. The performance analyzer subscribes to the sensor data from the sensor buffer and conducts the model inference of each application. In addition, it measures the performance metric and calculates

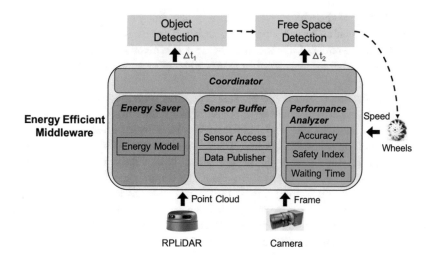

Fig. 4.4 E2M high-level overview

the maximum waiting time for each running application. Because each application has a specific objective, each application has a specific performance metric. For example, the performance of free space detection is determined by the safety of the robot. For object detection, the performance is determined by the accuracy in recognizing objects. Because each application has its own metric, in order to include a new application in E2M, the user (e.g., developer) just needs to input the performance metric, which will then be measured by the performance analyzer for E2M decisions. The energy saver automatically estimates the energy consumption of each application for various waiting times. All the waiting time, energy model, and performance analyzer results are sent to the coordinator, which determines how multiple applications could work together to maximize the energy saving. The speed of the AMR is an important input to E2M, and it affects not only the performance requirements of each application but also the energy saving of E2M.

4.1.3.1 Sensor Buffer

Uncoordinated access to sensor data is one of the inefficiency sources of energy of AMRs. In fact, multiple computer vision-based applications running simultaneously can share the sensor but have exclusive access to it to ensure reliability. The reason for this inefficiency is that the sensor driver gives exclusive access to the application reading the sensor. This means that multiple applications accessing the sensor within a short period of time may read similar data multiple times, thus wasting sensor's driver and data capturing energy. One approach to solve this problem is to sharing the data in the application level [600]. Another approach is to add a sensor buffer in the middle to share the data. Therefore, E2M implements a sensor buffer application to share the video frames among multiple applications to reduce the energy consumption.

The sharing of the video frames across concurrently executing applications is based on ROS's publisher/subscriber mechanism. The design is to capture the sensor data from camera and RPLiDAR and share the data with other ROS nodes to process the data. The size of the sensor buffer is statically defined and ROS manages to drop the extra data.

Each application subscribes to the data it requires from the sensor buffer. During the waiting time periods where no application needs data from a sensor, the sensor data will still be captured and published by the sensor buffer, but the data is not subscribed by other ROS nodes. There are two reasons for this design. First, although it is possible to stop the running sensor buffer, the sensor device does not support the operation of shutdown by receiving a command from the computing platform. Second, the waiting time is on the milliseconds level, while the time it takes to restart the sensor node and publish data out is on the seconds level. If the sensor is turned off, the data will be lost before it restarts. Therefore, the sensor buffer just keeps running in the design of E2M.

4.1.3.2   Performance Analyzer

E2M is designed to save energy with guaranteed performance. Thus, how to quantify the performance becomes an essential problem. In general, the performance metric must be defined based on the application. Because free space detection determines not only the direction but also the speed of the robot, the performance is determined by the safety of the robot. For object detection, the performance is determined by the detection accuracy. Thus, we define the cumulative accuracy for object detection to evaluate its performance. One problem is how to compare different performance metrics to make decisions about the coordinated waiting time of applications. In order to correctly compare different metrics, all the performance metrics are normalized using their desired value. Thus, for any application, a good performance is achieved with a normalized performance value near one. Next, we define two metrics to quantify the performance of the two example applications implemented on HydraOne, i.e., the safety index and the cumulative accuracy.

Safety Index In this chapter, we assume that if the robot does not crash into any object, then the control of the robot is safe. Based on this assumption, we propose a safety index that aims at keeping HydraOne from collision. As Fig. 4.5 shows, there are two cases of HydraOne's safety index computation. The linear (V) and angular (α) speed of the robot use polar coordinates with origin in the front of the robot. For cases 1 and 2, the idea is to calculate the safety index as the subtraction between the front nearest object distance and the braking distance. The robot's speed is decomposed into the front and the right directions of the robot. The AMR distinguishes case 1 with case 2 based on the point cloud data from RPLiDAR. The idea is to compare the change of distance around the degree with the shortest distance: if the shortest distance value is the trough, then E2M uses case 1; otherwise, it uses case 2.

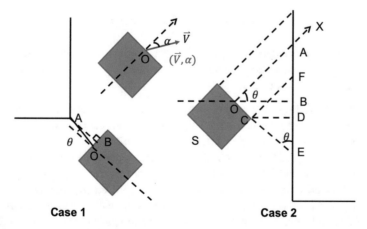

Fig. 4.5 Two cases of safety index

Case 1 The robot heads to a corner of the wall. In this case, the difference between the shortest distance to obstacles and braking distance gives the safety index for case 1, which is expressed as follows:

$$S_1(v(t), a(t)) = a(t) \cos \theta(t) - \left[v(t) \cos \alpha(t) t_0 + \frac{(v(t) \cos \alpha(t))^2}{2 \mu g} \right] \quad (4.1)$$

In Eq. 4.1, t is the time and $v(t)$ is the speed of the robot at time t. $S_1(v(t), a(t))$ represents the safety index of case 1 given the speed $v(t)$ and the shortest distance $a(t)$. The shortest distance OA is represented as $a(t)$ and the degree between OA and the front direction is denoted as $\theta(t)$. t_0 represents the reaction time and μ is the static fraction coefficient. μ reflects the relationship of static fraction with robot's weight and it is determined by the materials of wheels and ground.

Case 2 The robot heads toward a wall. In this case, the shortest distance that can be read from RPLiDAR's scan message is OB. Because the robot can only go forward, when reading the message from RPLiDAR, only those messages whose degree is within $-\pi/2$ to $\pi/2$ are used to find the nearest point. Both OB and the robot's speed are decomposed in the front direction and right direction of the robot. Then we can calculate the difference between shortest distance to obstacles and braking distance for each direction. Then the lower difference between the two directions is chosen as the safety index. Equations 4.2, 4.3, and 4.4 show how to calculate safety index for case 2.

$$d_x(t) = \frac{a(t)}{\cos \theta(t)} - \frac{s}{2} \tan \theta(t) - \left(v(t) \cos \alpha(t) t_0 + \frac{(v(t) \cos \alpha(t))^2}{2 \mu g} \right) \quad (4.2)$$

$$d_y(t) = \frac{a(t)}{\sin \theta(t)} - \frac{s}{2} - \left(v(t) \sin \alpha(t) t_0 + \frac{(v(t) \sin \alpha(t))^2}{2 \mu g} \right) \quad (4.3)$$

$$S_2(v(t), a(t)) = \min\{d_x(t), d_y(t)\} \quad (4.4)$$

The speed $v(t)$ is decomposed into x-axis (front) and y-axis (right). $d_x(t)$ represents the shortest distance in the x-axis at time t. $d_y(t)$ represents the shortest distance in the y-axis at time t.

Cumulative Accuracy For object detection, we use accuracy as performance metric. Currently, most related work solutions evaluate the performance of object detection algorithms by utilizing Intersection of Union (IoU) and mean Average Precision (mAP) [172]. However, for a trained object detection model (e.g., MobileNet-SSD in HydraOne), the detection accuracy for a specific frame does not change in different executions because we do not change the trained model. Assuming the robot has 0.1 s waiting time, if we compare the object detection results with waiting time and without waiting time, the average IoU and mAP will be the same because the DNN model is the same. However, objects during the waiting time are not analyzed by the model. As a result, the baseline that continuously runs without waiting time will have the highest number of objects recognized during a certain activity period. Therefore, we design the cumulative accuracy (*CA*) to measure how many objects are analyzed and recognized during a certain active period and evaluate the object detection performance:

$$CA = \sum_{i=1}^{N} \max \{a(i, j), j \in [1, M_j]\} \tag{4.5}$$

where N is the number of times the model inference has been executed. For the i-th model inference, $a(i, j)$ represents the output of the model inference, which is a vector consisting of M_j elements, and each element indicates the probability of being recognized as a particular object. Thus, $\max \{a(i, j), j \in [1, M_j]\}$, selects the recognized object with the highest probability for each model inference execution. As a result, CA is the sum of all the highest possibilities over N times of model inference.

Metric Normalization The performance metrics of different applications and their ranges can be different. The goal of normalization is to make E2M a general support for all computer vision-based applications in AMRs and to easily compare the performance of different applications.

In order for the coordinator to make a final decision on the waiting time of the applications, the coordinator needs to consider the performance of each application (see Sect. 4.1.3.4). Therefore, the performance metrics should be normalized. For safety index, first we collect the human driving dataset by using the joy stick to control HydraOne with no crash. The lower the safety index is, the higher is the possibility for a crash to happen. Then the normalized safety index can be calculated using the desired safety index, which is determined as the largest safety index when the robot is controlled by a human being. For *CA*, we also use desired *CA* value to divide the real-time *CA* value. The desired *CA* values are defined as the total number of objects through the detection process.

$$N_S = \frac{S(v(t), a(t))}{S_{des}} \tag{4.6}$$

$$N_{CA} = \frac{CA(v(t), b(t))}{CA_{des}} \tag{4.7}$$

In Eqs. 4.6 and 4.7, N_S and N_{CA} represent normalized safety index and normalized cumulative accuracy, respectively. The higher N_S is, the more safe is the AMR. The higher N_{CA} is, the better the AMR recognizes objects. Specifically, S_{des} and CA_{des} represent the desired safety index and cumulative accuracy, respectively. As a result, the performance analyzer also calculates the per-application waiting time. The maximum waiting time for HydraNet and MobileNet-SSD can be calculated based on the distance and speed as follows:

HydraNet Waiting Time

$$W_{AD}(v(t), a(t)) = \frac{S(v(t), a(t))}{v(t)} - t_{AD} \qquad (4.8)$$

MobileNet-SSD Waiting Time

$$W_{OD}(v(t), b(t)) = \frac{\min\{b(t)\}}{v(t)} - t_{OD} \qquad (4.9)$$

In Eqs. 4.8 and 4.9, $W_{AD}(v(t), a(t))$ and $W_{OD}(v(t), b(t))$ represent the maximum waiting time for HydraNet and MobileNet-SSD, respectively, given the speed $v(t)$ and distance information $a(t)$ and $b(t)$. The higher $W_{AD}(v(t), a(t))$ and $W_{OD}(v(t), b(t))$ are, the more energy can be saved. $a(t)$ represents the shortest distance to the RPLiDAR and t_{AD} is the average inference time for HydraNet. $b(t)$ is the distance to objects and t_{OD} is the average inference time for MobileNet-SSD. The distance to the objects is predicted based on the RPLiDAR's *scan* message and the object's location in the image.

4.1.3.3 Energy Saver

Energy saver is designed to determine the total energy consumption of the system when the waiting time of each application changes. The results from the energy saver are directly fed into the coordinator. Equations 4.10, 4.11, and 4.12 show the energy model before applying E2M:

$$E_{Baseline} = E_{other} + E_{AD} + E_{OD} \qquad (4.10)$$

$$E_{AD} = E_{AD-capture} + E_{AD-inference} + E_{AD-publish} \qquad (4.11)$$

$$E_{OD} = E_{OD-capture} + E_{OD-inference} + E_{OD-show} \qquad (4.12)$$

$E_{Baseline}$ is the total energy consumption when E2M is not applied. $E_{Baseline}$ consists of three components: E_{AD}, which is the energy consumption of running end-to-end free space detection application, i.e., HydraNet; E_{OD}, which stands for the energy consumed by running object detection application, i.e., MobileNet-SSD; and E_{other}, which represents the energy consumption of other parts, including sensors powering and wheels rotation. For each running application, HydraNet and MobileNet-SSD, we further break the energy consumption into sensor data capturing $E_{AD-capture}$ or $E_{OD-capture}$, model inference $E_{AD-inference}$ or $E_{OD-inference}$, and publishing commands/showing results $E_{AD-publish}$ or $E_{OD-show}$.

After applying E2M, the energy model can be represented as

$$E_{E2M} = E_{other} + E_{sensor-buffer} + E_{E2M-overhead}$$

$$+ \frac{t_{AD} - \sum W_{AD}}{t_{AD}}(E_{AD-inference} + E_{AD-publish}) \qquad (4.13)$$

$$+ \frac{t_{OD} - \sum W_{OD}}{t_{OD}}(E_{OD-inference} + E_{OD-publish})$$

From Eq. 4.13, we can see that the sensor buffer is introduced to capture sensor data instead of letting each application access the sensor directly. E_{E2M} represents the total energy consumption after applying E2M. $E_{sensor-buffer}$ represents the energy consumption of capturing and sharing sensor data. The length of waiting time of each application also affects the energy consumption of model inference and output. In addition, there is computation overhead in implementing the system: safety index, cumulative accuracy, and publisher/subscriber mechanism in ROS. Here we use $E_{E2M-overhead}$ to represent the total energy overhead.

4.1.3.4   Coordinator

The uncoordinated execution of concurrent jobs is another source for inefficiency on AMRs. The coordinator is proposed to coordinate multiple applications so that the overall energy consumption is minimized. Our design is shown in Fig. 4.6. We can see that there are two applications: application A and application B. Based on the energy and waiting time model we have discussed above, the maximum waiting time and energy saving for each application can be calculated. However, the problem is how to coordinate them to minimize the energy consumption while ensuring good performance. Our idea is to coordinate the waiting time and inference time to maximize the idle time. Indeed, increasing the idle time of computing resources helps the computing hardware to reach the deep-sleep state more often for energy savings.

In E2M, we assume the presence of a safety-related application simply because autonomous mobile robots need at least one navigation application like HydraNet to be executed. Navigation applications are often related to a safety index performance metric. As a result, the main role of the coordinator is how to coordinate the safety-related application with the other applications to minimize the energy consumption. Therefore, we enforce a restriction on their calculated waiting time. The maximum waiting time of the other applications will not exceed that of the navigation application. For example, if the computed waiting time of MobileNet-SSD is longer than the waiting time of HydraNet, we assign the waiting time of HydraNet to MobileNet-SSD. The reason for implementing this restriction is that the execution of HydraNet's inference changes the speed and direction of HydraOne, which indirectly affects also the performance and thus the waiting time of the other applications. As a result, the coordinator optimization problem is defined as follows:

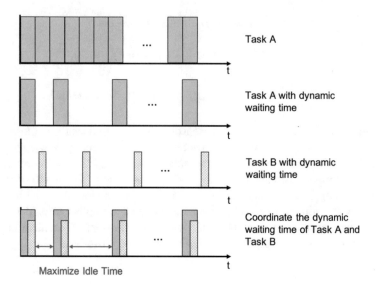

Fig. 4.6 The design of how E2M coordinates multiple applications

$$\textbf{min} \quad E_{E2M}$$

$$\textbf{s.t.} \quad 1 - \varepsilon_k < N_k < 1 + \varepsilon_k \qquad\qquad\qquad\qquad (4.14)$$

$$0 \leqslant W_k \leqslant \min(W_k^{\max}, W_S^{\max}) \quad \forall k \in [1, A]$$

For every coordinator activation, the objective is to minimize the energy consumption of the computing platform and the decision variables are the waiting times for each application for the current coordination period. Assuming there are A applications, for each application k there are three restrictions: the first is that the normalized performance metric of the application k should be close to the desired value, i.e., one, where ε_k is defined by the user and represents the acceptable error from the desired value; the second is that the waiting time for application k is positive and it is lower than the minimum between the maximum desired waiting time $W_S^{\max}$ of the navigation application S and the maximum desired waiting time $W_k^{\max}$ of the kth non-navigation application (e.g., as calculated for HydraNet and MobileNet-SSD in Eqs. 4.8 and 4.9, respectively).

When there is only one application running (i.e., the navigation application), the output waiting time of the sensor buffer is directly applied without the need of publishing/sharing results. Because the coordinator is needed when multiple applications concurrently execute, here we cover the case when there is more than one application running in parallel. In addition, because we do not predict performance analysis and energy savings over future periods, the coordinator's work is to find the optimal waiting time with guaranteed performance for the current coordination cycle. The optimization process consists of five main steps:

Step 1:   Wait for a new inference of the navigation application. When the navigation application finishes the current inference, its waiting time W_S is calculated. To minimize the energy consumption, by default the waiting times of all the applications are set to their max value (i.e., $N_k = 1$ $\forall k \in [1, A]$).

Step 2:   For the non-navigation applications with waiting time W_k higher than W_S, assign the waiting time of navigation application to these applications ($\forall k \in [1, A]$, s.t. $W_k > W_S$, then $W_k = W_S$). The reason is that the navigation application controls the speed and direction of the robot. If the other non-navigation applications wait more time than the navigation application, then the speed and direction of the robot change during their waiting time, which may degrade their performance.

Step 3:   For the non-navigation applications that have a shorter waiting time than the navigation application ($\exists k \in [1, A]$ *except S*, s.t. $W_k < W_S$), get the minimum waiting time W^{min} among these non-navigation applications;

Step 4:   After W^{min} has passed, execute all the model inference of non-navigation application whose waiting time is shorter than W_S.

Step 5:   After all the applications have finished their model inference, check if there is still enough time to wait for the new minimum waiting time plus the model inference time. If yes, go to Step 3. If not, go to Step 1.

Note that we use this heuristic algorithm to solve the above coordinator optimization problem. We plan to improve the design of the coordinator algorithm in our future work to provide theoretical guarantees of optimality.

4.1.4   Implementation

The implementation of E2M is conducted on HydraOne platform. The HydraOne platform is shown in Fig. 4.1b. HydraOne is an indoor autonomous mobile robot embedded with two leopard cameras and an RPLiDAR. The leopard camera has $1928H \times 1088V$ active pixels and a frame rate of 30 fps. The RPLiDAR's accurate range is from 0.2 to 12 m, which is enough for indoor scenarios. Each sample acquired by the RPLiDAR is composed of 8000 distance points around $360°$ view. The computing platform is an NVIDIA Jetson Tegra X2 board that has an NVIDIA $Pascal^{TM}$ Architecture GPU, 2 Denver 64-bit CPUs with Quad-Core A57 Complex, and an 8 GB L128 bit DDR4 Memory. The cameras and RPLiDAR are connected to the Jetson TX2 board via USB 3.0 port.

The ROS framework design of E2M is shown in Fig. 4.7. An ROS node (i.e., the nodes in the figure) is a process to perform a certain computation, while ROS topics (i.e., the arrows in the figure) are named buses for ROS nodes to exchange messages [443]. In E2M, we implement seven ROS nodes to access the sensor data and process the data. Six ROS topics are implemented to exchange messages including images, point clouds, control commands, and other customized messages.

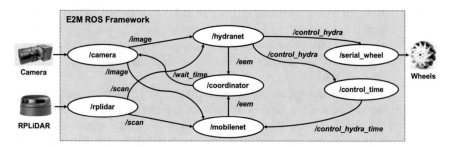

Fig. 4.7 The ROS framework of E2M

4.1.4.1 ROS Nodes

Each ROS node runs as a process to perform computation. Among the seven ROS nodes, */camera* and */rplidar* are used to access the sensors to capture and publish the data. Thus, the sensor buffer is implemented within these two nodes. The performance analyzer and energy saver are implemented in ROS nodes */hydranet* and */mobilenet* accordingly. The ROS node */hydranet* also executes the model inference of HydraNet for free space detection and calculates the safety index as well as waiting time for HydraNet. The ROS node */mobilenet* executes the model inference of MobileNet-SSD for object detection and measures the cumulative accuracy and waiting time. The ROS node */serial_wheel* is designed to get the control message */control_hydra* from */hydranet* and publishes them to the Arduino board that controls the motor drivers. */control_hydra* does not have timestamp, but it is necessary for ROS node */mobilenet* to synchronize the messages, so the ROS node */control_time* is introduced to convert the control message */control_hydra* to a timestamped control message */control_hydra_time*. The */coordinator* subscribes to messages including waiting time, performance analysis, and energy saving to calculate the coordinated waiting time and then publishes it to */camera*.

4.1.4.2 ROS Topics

The ROS topics are defined to exchange messages between ROS nodes. There are six ROS topics in E2M. The summarized description of these six topics is reported in Table 4.2. The ROS topics */image* and */scan* belong to */sensor_msgs*, while */control_hydra* and */control_hydra_time* belong to the */geometry_msgs* library. The header of each ROS topic includes the timestamp of the message. The image pixels are stored in the field "data" of */image*. The distance message is stored in the field *ranges* of */scan*. The ROS topic *control_hydra_time* is generated by the ROS node *control_time*, which subscribers to the *control_hydra* message and adds to it the header with the timestamp for */mobilenet*.

Two customized ROS messages */eem* and */wait_time* are defined based on *String* type in */std_msgs*. */eem* is defined for ROS node */hydranet* and */mobilenet* to share

Table 4.2 ROS topic description in E2M

Topics	/image	/scan	/control_hydra	/control_hydra_time	/eem	/wait_time
Library	Sensor_ msgs	Sensor_msgs	Geometry_ msgs	Geometry_ msgs	Std_ msgs	Std_msgs
Type	Image	LaserScan	Twist	TwistStamped	String	String
Fields	Header, height, width, encoding, data, etc.	Header, angle_min, angle_max, angle_increment, ranges, intensities, etc.	Linear, angular	Header, linear, angular	Header, wait_ad, wait_od, safety, accuracy	Header, time

some real-time information including waiting time, safety index, and accuracy with ROS node */coordinator*. */wait_time* is defined to send information of coordinated waiting time to the sensor buffer */image*.

4.1.4.3   Message Synchronization

The synchronization issue arises when one application needs to process data from different sensors. In E2M, each ROS node is an individual process that can run anytime, so they can all run in parallel. The ROS topics are used for ROS nodes to share message. However, as the communication delay is a random process, the messages published at the same time may not arrive at the subscribers at the same time. For ROS node that subscribes to multiple ROS topics and processes the messages together, the synchronization of different ROS topics becomes an essential problem. For example, both ROS nodes */hydranet* and */mobilenet* subscribe to topics */image* and */scan*. For distributed systems, the synchronization is usually based on time. Thus, in order to synchronize multiple ROS topics, we first need to attach timestamp to all the ROS topics. Therefore, we add an ROS node */control_time* to transform the control message to a timestamped control message.

When an ROS node subscribes to ROS topics, the node uses a callback mechanism to get the real-time message. When multiple ROS topics are subscribed, the callback function should be called after the synchronization of all the ROS messages. In E2M, the synchronization of multiple ROS topics is implemented based on *message_filters*. We use the *ApproximateTime* policy with buffer length equal to 10 and the slop equal to 0.05 s. The buffer is used to store the unsynchronized messages, while the slop defines the maximum timestamp difference for synchronization.

4.1.5   Discussion

Although E2M achieves nearly 24% of energy saving with less than 8% performance degradation, there are still some limitations owing to the design and
targeted scenario. In this section, we discuss the limitations of E2M in three aspects:
generality of E2M, multiple applications coordination, and random process of model
inference.

4.1.5.1   Generality of E2M

E2M is a middleware designed mainly for autonomous mobile robots used for
industry purposes. E2M is able to be deployed in any ROS-based AMRs. New
applications can be implemented as plugins. For example, if we want to add the
application of avoiding unexpected obstacles, we can change the way the performance metric for the navigation application is calculated. According to [254], we
can introduce a distance space metric, which suggests that AMR applications need
to respond within a certain response time given by Response_space(m)/speed(m/s).
For example, if we want a new decision every 1 m at most and the current speed is
40 km/h, then the desired response time is 90 ms. We could use this as performance
metric, without changing the described E2M algorithm.

For real autonomous vehicles, they have several commonalities with AMRs,
including the need of analyzing camera images for object detection and driving.
However, real environment is so complicated that even the most complex deep
learning model may fail. Hence, the waiting time of the model inference may cause
a considerable performance degradation. In addition to this, the multiple cameras
on-board with multiple applications running in parallel, the coordination of sensor
access and the coordination of the application executions can contribute to achieve
energy savings even in the autonomous vehicle case.

4.1.5.2   Multiple Applications Coordination

In this chapter, we only consider the coordination of concurrent applications in
one cycle, which makes the case become easier because the coordination is just a
local optimal solution. However, if we want to find the global optimal solution that
achieves the most energy saving with guaranteed performance, we need to predict
the performance changing, which can be very challenging. In addition, as we have
discussed in the performance analysis of the model inference latency, ROS node's
messages will contribute to the delay of the message. Finally, the information that
the coordinator gets can be all out of date. These problems may make it hard for
E2M to save energy while guaranteeing performance. One solution is to use the
newer ROS 2.0 rather than the currently used ROS, which introduces shared memory
to improve the throughput and latency of data-sensitive communications.

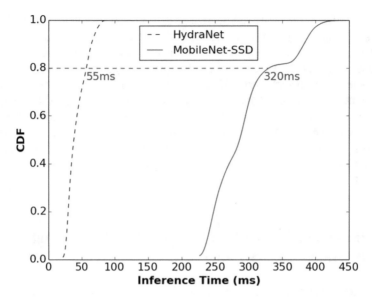

Fig. 4.8 The inference time CDF of HydraNet and MobileNet-SSD

4.1.5.3   Random Process of Model Inference

From Fig. 4.8, we can observe that the inference time is a random process. Although the model is fixed and the input images are all from the same scenario, the inference time can vary from 21 to 88 ms for HydraNet and 225–434 ms for MobileNet-SSD. The random process of model inference time makes the waiting time of each application become a random process. This randomness makes it difficult for us to predict how long the model inference and the waiting time can be. Hence, the inference time and the waiting time of each application also need to be dynamic, which makes the coordination of concurrent applications challenging.

4.1.6   Summary

Autonomous mobile robots (AMRs) have been widely utilized in industry to execute various on-board computer vision applications. Most of the applications involve the analysis of camera images through trained deep learning models. In this chapter, we have first analyzed the breakdown of energy consumption for the execution of computer vision applications on AMRs and discovered three main root causes of energy inefficiency: uncoordinated access to sensor data, performance-oriented model inference execution, and uncoordinated execution of concurrent jobs. In order to fix these three inefficiencies, we have proposed E2M, an energy-efficient middleware software stack for autonomous mobile robots. First, E2M regulates the access of different processes to sensor data, e.g., camera frames, so that the amount of data actually captured by concurrently executing jobs can be minimized. Second, based on a predefined per-process performance metric (e.g., safety, accuracy) and desired target, E2M manipulates the process execution period to find the best energy-performance trade-off. Third, E2M coordinates the execution of the concurrent processes to maximize the total contiguous sleep time of the computing hardware for maximized energy savings. We have implemented a prototype of E2M on a real-world AMR. Our experimental results show that E2M leads to 24% energy savings for the computing platform, which translates into an extra 11.5% of battery time and 14 extra minutes of robot runtime, with a performance degradation lower than 7.9% for safety and 1.84% for accuracy.

4.2   Determinism Analysis of Deep Neural Network Inference for Autonomous Driving

4.2.1   Introduction

Owing to its high safety and efficiency, autonomous driving has become the fundamental technology for the next generation of transportation. Deep neural networks are widely deployed in the autonomous driving system for sensing, perception, decision, and control tasks. Typical examples include YOLOv3, Faster R-CNN for object detection [436, 441]; MobileNetv2, Deeplabv3 for image semantic segmentation [76, 222]; and LaneNet and PINet for lane detection [265, 380]. There are two main reasons for the success of DNNs in autonomous driving systems. The first is that they achieve higher accuracy compared with traditional vision-based approaches [184]. The other one is that DNNs can process raw data, which makes it suitable for autonomous driving vehicles since there are many sensors on the vehicles generating terabytes of raw sensor data every day [310].

As a safety-critical system, autonomous driving gives high requirements in accuracy, real-time, robustness, etc. The improvement of accuracy by DNN-based algorithms promotes the development of the autonomous vehicle, but how to satisfy the real-time requirements of the sensing, perception, and decision tasks is still a significant challenge. Typically, a massive amount of sensor data can be generated and processed by the on-broad computing system in real-time. According to [254], when the vehicle drives at 40 km per hour in urban areas and that autonomous functions should be effective every 1 m, each real-time task's execution should be less than 100 ms. As DNN models are widely used in object detection/classification, lane tracking, and decision-making applications, how to guarantee the real-time execution of the DNN inference becomes the key to satisfy the real-time requirements of safety-critical tasks.

Generally, for safety-critical applications like sensing, perception, control, etc., the deadline will be used by the real-time scheduler to guarantee safety. The setting up of deadlines is usually based on the worst-case execution time. However, although the model structure and weights are fixed, we can still observe huge time variations in DNN inference execution. It brings a big challenge for real-time scheduler because the deadline with worst-case execution time could waste many processor resources. If the DNN inference time could be fixed or narrow down to a specific range, many resources would be saved. Therefore, *analyzing the determinism of the DNN inference becomes the key to optimizing DNN inference runtime*. However, most of the works on deterministic either focus on code level or enable the schedule to guarantee determinism [154, 480]. A detailed and in-depth analysis of determinism of DNN inference is missing.

In this chapter, we undertake a comprehensive analysis of DNN inference's determinism in a typical autonomous driving system [311]. We analyze the time variation issues for typical model granularity as well as end-to-end system granularity. For a typical model, we consider the variability in DNN inference from five perspectives: *data, I/O, model, runtime, and* hardware. For the profiling of system granularity, an end-to-end system is built based on ROS (Robotics Operating System) to reveal real autonomous vehicles' time variations. Based on the analysis of the determinism, six findings are summarized for achieving deterministic DNN inference.

In summary, this work makes the following contributions:

- The time variation issues in DNN inference for autonomous driving are thoroughly studied. We found that the majority of DNN models show variations larger than 100 ms, which significantly affects autonomous driving safety.
- Through a comprehensive analysis of the determinism of a typical DNN model inference from five perspectives, we observed several findings in the relationship between time variation and variability of data, I/O, model, runtime, and hardware.
- We observed that the variation goes to seconds level through the system-level analysis on an end-to-end autonomous driving system and summarized several suggestions for building a deterministic DNN inference system for autonomous driving.

4.2.2   Time Variation in DNN Inference

The proliferation of deep learning achieves enormous performance improvement and brings lots of issues in computation complexity, energy consumption, and the determinism for safety-critical systems [564]. Generally, time variation reflects the determinism of the system. For autonomous driving vehicles, determinism makes predictivity, which guarantees the vehicle's safety and improves resource utilization [404].

However, current DNN-based computing systems show poor performance in terms of determinism [309]. To illustrate the time variation issues of the state-of-the-art DNN-based autonomous driving system, we choose seven models used in various applications and measure the end-to-end latency with

Table 4.3 The mean, range, and variations of the seven DNN models used in the autonomous driving system

Model	Mean (ms)	Range (ms)	Variation (%)
YOLOv3	174	57	32.8
Faster R-CNN Resnet101	413	**128**	31
Mask R-CNN Inceptionv2	266	104	39.1
SSD-MobileNetv2	144	70	48.6
PINet	127	**263**	**207.1**
LaneNet	82	**282**	**344**
Deeplabv3 + MobileNetv2	149	19	12.8

Bold values are larger than 100, which means the time variations are large than 100 ms

the same input images. Among all the seven DNN models, YOLOv3 [436], Faster R-CNN ResNet101 [441], Mask R-CNN Inceptionv2 [200], and SSD MobileNetv2 [222, 320] are for object detection, PINet [265] and LaneNet [380] for lane detection, and Deeplabv3 + MobileNetv2 for semantic segmentation. Table 4.3 shows the results of the mean, the range, and the variation percentage of the end-to-end latency. The range is defined as the difference between the maximum value and the minimum value. The variation percentage is defined as the range divided by the mean value. Table 4.3 shows that four of the seven DNN models have a range larger than 100 ms. If we consider the percentage, the variations of the last three models for lade detection and semantic segmentation are all larger than 100%. LaneNet shows the poorest performance with 344%.

How about the distributions of the end-to-end latency? Figure 4.9 shows the CDF of the end-to-end latency of the above seven models. We can observe that the latency distribution is normal, and there is a non-negligible amount of latency values that lie above 90% and even the 99% line. Although having such a huge amount of variation in end-to-end latency does not affect the algorithms' performance, it affects the scheduler's performance, especially for the safety-critical system like autonomous driving. The reasons are owing to the deadlines and penalties caused by missing deadlines. If we consider the autonomous driving system, a hard real-time system, the scheduler will assign deadlines for each task based on their history execution time. Take PINet as an example. It would assign the deadline larger than the worst-case execution time, which is the highest end-to-end latency 357 ms as the deadline [559]. This design guarantees the safety of this application. However, it also brings huge inefficiency because the real execution time is less than 185 ms over 90% of the time, which means over 170 ms may be wasted for most jobs. However, if we can narrow the range execution time to less than 50 ms, we can save almost 120 ms in PINet for every job.

Determinism is expected to make a significant impact on safety-critical real-time systems. Since the analysis of time variation varies with different applications and scenarios, this work's focus would be the DNN models in the autonomous driving system. We propose using profiling tools to observe variation issues existing in current autonomous driving systems and provide suggestions for building deterministic systems.

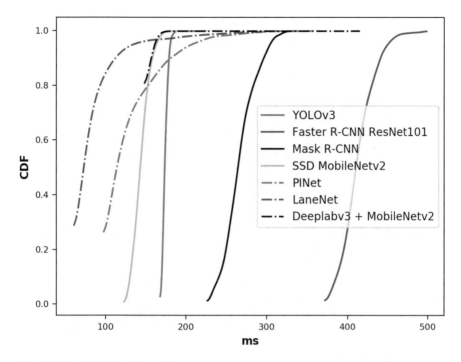

Fig. 4.9 The CDF of end-to-end latency

4.2.3 Uncertainties in DNN Inference

What are the potential issues affected the determinism of DNN inference? To answer this question, we first analyze the timeline of DNN inference. Figure 4.10 shows a typical DNN inference time in TensorFlow. Assuming the whole pipeline starts by calling *inference()*, the graphs and weights will first be loaded into the memory. The process will then start reading the input; we are using *imread()* in OpenCV to read the image from a local path. Next, it will do some pre-processing on the image, including resizing, converting color space from one to another, etc. Next, the processed image will be passed to the *Session* and loaded together into the processor to run inference. Finally, some post-processing works like adding bounding boxes (*vis_util.box()*) will be called to append objects, lanes, etc. to the image.

From the timeline of DNN inference, we can find several uncertainties that might contribute to the time variation issue. The first is data variation, which means the value and distribution of pixels vary from image to image. The sparsity matrix is expected to have less time on inference than the dense matrix [82]. The second is on data's I/O to the session's graph. How the running graph reads data from the camera, image file, or through an ROS message could affect the inference's variation. The third is the model variation, which means the model's complexity in multiply and accumulate (MAC) operations. The more MACs a model has,

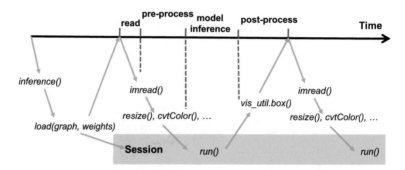

Fig. 4.10   The timeline of DNN inference

the more diversity in inference time. The fourth is runtime variation, owing to the competition of concurrent jobs for resources like memory, CPU, and GPU. How many processes allow preemption, their scheduling policies, priorities, etc. is the question that affects the process's runtime performance. Finally, the hardware variation also affects the time variation of DNN inference. GPU is expected to run faster than CPU, but a multi-core system is supposed to show more variation than a single-core system. How will different architectures affect the deterministic of DNN inference? In summary, there are five aspects of uncertainties in DNN inference: data, I/O, model, runtime, and hardware.

4.2.4   Profiling Tools

For current DNN-based autonomous driving systems, a big challenge is how to explore the determinism in DNN model inference. Our approach is profiling the system with a variety of granaries. In general, we use three profiling tools in this work: application level (*TensorFlow Profiler*), code level (*Python cProfiler*), and Linux system call level (*Linux Perf*).

TensorFlow Profiler   In TensorFlow [1], the execution of model inference is implemented as a call of a function graph. Each node in the graph defines a specific operation. *TensorFlow Profiler* is an application-level tool for analyzing the code to get the execution time and resource consumption of each node in the graph. We use *TensorFlow Profiler* to get the timelines and flow graph of the DNN inference.

Python cProfiler   *Python cProfiler*[1] is a library provided in Python for code-level profiling. It collects statistics that describe how often and for how long parts of the program are executed. The number of function calls can identify bugs in code and identify possible inline-expansion points (high call counts). Internal time statistics

---

[1] https://docs.python.org/3.8/library/profile.html#module-cProfile.

can be used to identify "hot loops" that should be carefully optimized. Cumulative time statistics should be used to identify high-level errors in the selection of algorithms. We use *Python cProfiler* to get the call graph with time breakdowns of the code.

Linux Perf Although application-level profiling gives us some explanations of model's variations, it is not enough for us to explain the variability of the model inference time for different architectures. Therefore, we conduct system call-level profiling with *Linux Perf* to show model inference performance in system call and operating system level. *Linux Perf* used the Linux system performance counter to monitor the whole Linux system [183]. We use *Linux Perf* to collect the system-level metrics, including CPU utilization, contest switches, CPU migrations, page faults, cycles, instructions, branches, branch misses, cache misses, etc.

Chapter 5
Dataset and Benchmark

5.1 Open Dataset for Autonomous Driving

5.1.1 Introduction

As the cornerstone of autonomous driving research, dataset makes researchers avoid the time-consuming, labor-intensive, and cost-intensive process of building autonomous vehicles to carry out experiments. Research based on datasets is a common research route for most autonomous driving research institute [310]. In the past decade, a batch of classic datasets [62, 94, 163, 447] have greatly promoted the development of autonomous driving research, and L3 autonomous driving technology has become mature in conventional scenarios. However, faced with more complex and diverse L4 and L5 autonomous driving tasks, traditional datasets are increasingly unable to meet the requirements of reliability redundancy in autonomous driving research. In recent years, datasets have developed in the direction of larger data, more scenes, more sensor combinations, higher quality images, higher quality labels, unstructured scenes, end to end, and research on driver behavior identification, etc.

The dataset providers and research users have different concerns about the dataset. Although data providers often collect data for one or several specific tasks when providing datasets, they pay more attention to the equipment and data itself, such as equipment selection and assembly, data types and quantities, label types and quality, scenario coverage and complexity, etc., they strive to make the dataset highlighter than previous datasets in some aspects. However, less of them focus on whether the data is more suitable for these tasks research. In the comparison, dataset users pay more attention to autonomous driving tasks, which datasets can be used, what are their advantages, and which datasets are most suitable for research. The considerations involved include data types, label types and quality, whether code is available, whether to provide training/testing set, benchmark, and commonly used

Table 5.1 Comparison of AV dataset-related survey papers

| Related work | Time | Survey coverage | | | |
		Num_of_datasets	Datasets	Tasks	Algorithms
[583]	2017	27	✓	–	–
[191]	2019	45	✓	✓	–
[245]	2019	37	✓	✓	–
[589]	2020	18	✓	–	–
Ours	2021	50	✓	✓	✓

algorithms for this task. To bridge the gap of the different concerns of these two parties by investigating and analyzing is the main motivation of our work.

According to the survey, the Google Scholar search results of "autonomous driving dataset" have increased by nearly 9 times than that of 10 years ago, and the search index in 2020 is more than twice that of 3 years ago. Considering the rapid increasing of autonomous driving datasets, we only make a comparative analysis of surveys in recent years, as shown in Table 5.1. Yin et al. compared 27 datasets in 2017 [583] and 37 datasets in 2019 [245], respectively. They used Google and Snowball to analyze available datasets in the top 200 search results and compared them in detail and give the highlight of datasets. Guo et al. [191] sorted out a comparative analysis of 45 datasets from multiple dimensions, for example, data provided, annotation, traffic, diversity, etc. They aimed to introduce the influencing factors of driving performance, provided a complex comparison table. But most of datasets discussed are before 2018 and there is no further detailed analysis about the correspondence between data and tasks. In 2020, Yurtsever et al. [589] discussed common practices and emerging technologies of autonomous driving. They briefly compared 18 datasets in recent years, but there is no more detail being discussed. After all, surveys are mainly focused on the comparison of datasets, none of them above provide a good answer to the mapping relationship between datasets and tasks. Facing these issues, we investigated and analyzed 50 influential datasets, nearly half of them were published in recent year, and we aim to bridge the gap between data providers' and data users' different main concerns. And we conservatively believe that the main contributions of our work can be summarized as follows:

1. The first survey compares 50 datasets covering influential 24 datasets published in the past 3 years.
2. The first survey of algorithm guidance and dataset about the task of automatic driving with dataset.
3. The first survey analyzes the detail mapping relationship between autonomous driving datasets and tasks, which can be a reference for autonomous driving new hands and researchers to exchange their dataset to the most suitable one.

5.1.2   Object Detection

Accurate and real-time object detection regardless of weather conditions is crucial for the real-world application of autonomous driving. The camera has the advantages of higher resolution, stronger recognition, and lower price, which is often used for 2D environment perception. At the same time, LiDAR can obtain high-precision depth information for the construction of 3D scene understanding. Limited by the paper length, we just introduce the main algorithms around these two devices.

5.1.2.1   Data Description and Recommendation

The publicly available datasets have played a key role in pushing forward the development in many computed vision-based image analysis tasks by providing problem-specific examples and the corresponding ground truth [101, 307]. In the context of autonomous driving, the studies of [3, 64, 163, 501] have provided benchmarks to evaluate diverse algorithms on the same data and contributed to closing the gap between the laboratory testing environment and various real-world problems. We discuss some popular benchmarks, as are shown in Table 5.2.

As the most important perception task, object detection is the base of other tasks such as tracking and prediction. For single target detection, a batch of specific benchmark appeared, such as pedestrian dataset Caltech [116], bicycle dataset TDCB [301], traffic light dataset DTLD [150] and BSTL [39], road signs datasets GTSRB [220] in Germany and TT100K [614] in China, road damage dataset [339], obstacles detection dataset LostAndFound [415], accident and fire dataset Traffic-Net [393], etc.

Regarding the task of multi-object recognition, many datasets are different from the selection of equipment, the types of data generated, the labeling method, and the scene selection. Take the most influential dataset KITTI as an example. It uses a camera/LiDAR device combination method to collect and organize daytime travel images, videos, and LiDAR data in an urban scene and compare 15,000 frames and its corresponding LiDAR data made 2D/3D label annotation. The KITTI dataset is used to carry out deep learning-related tasks based on 2D/3D recognition, tracking, and semantic segmentation.

Many datasets are different from KITTI in all aspects. From the perspective of equipment selection, DECD [431] uses event camera to improve the accuracy of object change recognition, PandaSet [468] uses the latest PandarGT/Pandar64 LiDAR to provide 3D scene recognition, Oxford Radar RobotCar Dataset [34] and Astyx Dataset HiRes2019 [354] use radar as the main equipment to collect data, while nuScenes [64] is the first dataset to carry the full autonomous vehicle sensor suite. From the perspective of the types and quantity of collected data, nuScenes and Waymo collect more than 1M images, while D2-city, BDD100k, and lyft L5 provide video clips more than 10K. From the point of view of the label, KITTI and Waymo provide 2D/3D/LiDAR annotation, while nuScenes focus on image 3D

Table 5.2 Overview of datasets and tasks. Here, CO, UR, RU, HI, WE, SE, NI, IL, OD, OT, SS, LO, ST, Advanced, E2E, BE represent Code, Urban, Rural, Highway, Season, Night, Illumination, Object Detection, Object Tracking, Semantic Segmentation, Location, Stereo and Advanced Intelligent Service, respectively. Besides, "-" means no label but suitable; "~" means unknown; "0" means training/testing data; "*" means providing benchmark; and "√" means suitable

Datasets	Overview			Diversity								Tasks					
	Time	Cited	Scale	CO	UR	RU	HI	WE	SE	NI	IL	OD	OT	SS	LO	ST	Advanced
Caltech Ped [116]	2009	1251	11 GB	√	√								√0*	√			
GTSRB [220]	2011	454	1.6 GB	√	√							√0*					
TT100k [614]	2016	319	105.9G	√	√	√						√0*					
TDCB [301]	2016	74	44G	√	√							√0					
DTLD [150]	2016	13	a/n	√	√							√0	√				
BSTL [39]	2017	114	34.3 GB		√	√		√			√	√0*					
Road Damage [339]	2018	89	2.4 GB	√	√	√		√				√0*					
Traffic-Net [393]	2015	22	48 MB		√		√					√0					
LostAndFound [415]	2016	36	48 GB		√							√0					
TRoM [321]	2017	4	100 MB	√	√							√0*					
Boxy [38]	2019	9	1.1 TB	√			√			√		√0*					
CADC [416]	2020	12	92 GB	√	√				√			√					
PandaSet [468]	2019	2	42G	√	√		√					√		√			
DECD [431]	2019	40	120G	√	√		√					~	~				
KITTI [163]	2012	5917	1.5h	√	√		√				√	√0* (IM/Li)	√0* (IM/Li)	√0* (IM/Li)	√0*	√0*	

Dataset	Year	#	Size												
nuScenes [64]	2019	382	400 GB	√	√		√			√	√0*(IM/Li)	√0*(IM/Li)	√0*(Li)		
UAVDT [121]	2018	139	10G			√	√			√	√			
Astyx HiRes2019 [354]	2019	23	350 MB	√						?	?	?	?	
Oxford Radar [34]	2019	42	4.7 TB	√	√				?	?	?			
BoxCars [489]	2018	60	6G	√	√				√	√0*(IM/Li)				
Stanford Track [510]	2010	184	2 GB	√	√					√0*(IM/Li)				
MOT16 [356]	2016	661	1.9G	√	√		√		√	√0*(IM/Li)	√0*			
Argoverse [73]	2019	177	300 GB	√	√		√		√	√	√0			
D2-city [74]	2019	13	>1 TB		√	√	√		√	√	√			
H3D [406]	2019	52	0.77h		√		√		√	√	√			
Waymo [501]	2019	121	2 TB	√	√		√		√	√0*(IM/Li)	√0*(IM/Li)			
Camvid [62]	2008	784	8 GB	√	√				√	√0*(IM/Li)		√		
Cityscapes [94]	2016	4088	63 GB	√	√	√						√0*		
Vistas [379]	2017	413	10 GB	√	√	√	√					√0*		
TuSimple [526]	2017	–	25G	√				√				√0*	√0(IM)	
The ApolloScape dataset [225]	2018	231	1.3 TB		√	√	√		√	√0*(Li)	√0*(Li)	√0*(Li)	√0*	√0*
BDD100K [586]	2018	56	1.8 TB	√	√	√	√		√	√0	√0*	√0		
IDD [535]	2019	43	18.5 GB	√	√	√			√	√0*		√0*		

(continued)

Table 5.2 (continued)

Datasets	Overview			Diversity								Tasks					
	Time	Cited	Scale	CO	UR	RU	HI	WE	SE	NI	IL	OD	OT	SS	LO	ST	Advanced
Highway Driving [261]	2020	3	644 MB				√							√0			
comma.ai [458]	2016	182	80G	√	√	√	√					√	√				E2E(√)
Udacity [527]	2016	–	300 GB	√			√	√	√			√0	√0		√0*		E2E(√/0*)
DBNet [81]	2018	55	>1TB	√	√		√	√			√	√					E2E(√/0)
HDD [426]	2018	84	104h		√	√	√				√	√					BE(√/0)
JAAD [276]	2016	45	3.1 GB		√			√		√		√	√				BE(√)
Brain4Cars [236]	2015	165	1180mi	√	√	√	√				√						BE(√/0*)
DR(eye)VE [8]	2016	58	35G	√	√	√	√	√		√							BE(√/0)
lyft L5 [221]	2019	7	100GB									?	?	?	?	?	
Ford Campus [402]	2009	241	200 GB	√	√							?	?	?	?	?	
Oxford RobotCar [338]	2017	588	23 TB	√	√			√	√	√	√	?	?	?	?	?	
Ford Multi-AV [3]	2020	7	>10TB	√	√		√		√								Multi-AV
A2D2 [167]	2020	20	2.3 TB	√	√							√					
UAH-DriveSet [445]	2016	75	3.3 GB	√			√							√			BE(√)
VKITTI [155]	2016	589	470 MB		√	√	√	√		√	√	√					Synthetic
SYNTHIA [447]	2016	952	>1T		√	√	√	√	√	√	√			√0			Synthetic
Apollo Synthetic [16]	2019	–	600 GB		√	√	√	√		√		√					Synthetic

detection benchmark, and label 40K frames with 3D bounding box and label 79M LiDAR point. CADC, pandaset, Astyx Dataset HiRes2019, and CADC pay more attention to 3D tags. In terms of scenarios, CADC's first wintry-related dataset CADC provides 75 snow scenes in the Waterloo area, Waymo [501] focuses on more diverse in collection scenarios, such as weather, seasons, night, illumination, etc., and IDD [535] is a novel dataset for unstructured environments.

Synthetic In the real world, collecting complex and diverse data is expensive and time-consuming, laborious, and error-prone. Synthetic solves these problems by simulating scenes with the game engine and only needs to capture data from the simulated vehicle. Almost all scene data can be collected, such as different weather conditions, different seasons, different times a day, and various traffic obstacles. Besides, the data collected are all high-precision ground truth.

In 2016, VKITTI [155] dataset simulated 5 urban scenes under different weather conditions and generated several types of ground truth, such as 2D/3D bounding box, semantic segmentation, etc. In the same year, SYNTHIA [447] dataset was released, which is focused on semantic segmentation task. It provides 10 times as many labeled frames as VKITTI and richer scenarios, including urban/highway/-green area scenes under different weather conditions/seasons/times a day. Apollo Synthetic [16] focus on 2D/3D box and semantic segmentation. It supplements indoor parking garage scenes and 3D lane line ground truth, which contains lane line degradation.

In addition, there are several important large-scale datasets. Only at the beginning of the release, the author made a small-scale mark and tested the detection performance. These large datasets have excellent characteristics in terms of data volume and related scenarios. These data have huge application prospects in the future after planned marking. For example, the Oxford RobotCar dataset [338] provides 100 repetitions of a consistent route, captured over a period of over a year, weather, traffic and pedestrian, the data volume reached 23.15 TB. BDD100K dataset [586] has 100K videos, each video is about 40 s long and with a resolution of 1280×720 pixels at 30 frames per second. And labeled keyframe images are extracted from the videos at 10th second. 100K high-definition videos over 1100 h of driving experience across different times in a day and weather conditions.

5.1.3 Object Tracking

5.1.3.1 Algorithms

Camera-Based Algorithms The solution of object tracking could be broadly divided into two phases: (1) detecting phase and (2) tracking phase. During the detecting phase, the state-of-the-art object detection algorithm, such as YOLOv3 (You-Only-Look-Once-v3) [436], a fast and accurate detector to detect targets and compute bounding boxes, so that each target is enclosed in a corresponding bounding box for each frame. Then, Deep SORT algorithm [561, 562], an extension to SORT [45] that incorporates deep appearance descriptors to reduce the number of identity switches caused by the long periods of occlusions. When a new detection is associated with a tracked target, the newly detected bounding box is used to update the current state of the target.

The Mahalanobis distance gives the information of targets' possible locations from the motion aspect, which is particularly effective for short-term prediction and tracking. Besides, the cosine distance considers appearance information, which is particularly useful to reduce identity switches caused by the long-term occlusions, especially when motion is less discriminative. Consequently, Deep SORT combines the Mahalanobis distances and the cosine distance as an indicator to figure out whether a frame-by-frame association is admissible to keep tracking multiple targets.

LiDAR-Based Algorithms Tracking is based on the estimation of the object position of objects in subsequent frames based on a given frame of data. LiDAR-based target tracking can be traced back to the introduction of Siamese network into 3D object tracking by Giancola et al. [170]. The method uses the Kalman filter to generate candidate objects in two steps, then generates a compact representation of the shape regularization through the coding model, and then matches the detected objects through cosine similarity. Qi et al. [422] proposed P2B (point-to-box method) for 3D target tracking. This method uses the PointNet++ backbone to generate the template and the seeds of the search area and also uses the target clues in the template to enrich the seeds of the search area. Then the target is proposed and verified, the candidate target center is returned, and the seed target is jointly proposed and verified.

5.1.3.2   Data Description and Recommendation

MOT16 [356] is an important dataset for studying multi-target detection, providing 14 sequences (7 for training and 7 for test), with 110,407 training boxes and 18,2326 test boxes, but MOT16 is not specifically collected for autonomous driving with only two sequences collected from bus. The KITTI dataset provides 2D MOT with object tracking benchmark consist of 21 training sequences and 29 test sequences collected from car, in which just only two classes "Car" and "Pedestrian" are used for tracking. In contrast, nuScenes [64] and Argoverse [73] provide 3D Multi-Object Tracking (MOT), and the former monitors 17,081 pieces of 7 types of information and the latter 15 types to 11,052 pieces of track data. From the top view, UAVDT [121] provides 2700 tracking data. The all datasets mentioned provide benchmarks.

LiDAR object tracking datasets commonly include KITTI [163], The ApolloScape dataset [225], Waymo Open [501], nuScenes [64], and H3D [406]. Among them, there are 917 trajectories tracked by KITTI, 13,763 in H3D and 64,386 in nuScenes, and Waymo has reached around 113k unique LiDAR tracking ID. The LiDAR datasets of KITTI, the ApolloScape dataset, and H3D are all collected in urban scenes, and there is no scene change. The nuScenes dataset adds weather changes, while Waymo has all the scenes.

5.1.4 Semantic Segmentation

5.1.4.1 Algorithms

The environment perception ability is supported by the rapid development of semantic segmentation technologies, and it is the fundamental of autonomous driving, while lane detection is one of the key parts of environment perception.

Camera-Based Algorithms Previous mainstream lane detection methods (e.g., [55, 508, 565]) rely on a combination of handcrafted features and heuristics to identify lane segments, and in general, they are prone to robustness issues due to road scene variations. Popular choices of such handcrafted cues include color-based features [85], the structure tensor [323], the bar filter [511], ridge features [324], etc., which are possibly combined with a Hough transform [308, 611] and particle or Kalman filters [264, 511]. After identifying the lane segments, post-processing techniques are employed to filter out errors and group segments together to form the final lanes [208].

LiDAR-Based Algorithms Similar to image semantic extraction, there are two main types of information extraction from LiDAR point cloud: semantic segmentation and instance segmentation. In semantic segmentation, a projection-based method is used to process three-dimensional point cloud data in two dimensions. For example, [295] projects the point cloud onto a cylindrical surface. The 3D point cloud is directly converted into a 2D projection. This type of method reduces the amount of calculation but loses part of the data. In contrast, the point-based method directly processes the 3D point cloud, which avoids the loss of information, but the computational cost is higher. This type of method attempts to reduce the information loss in the feature extraction process. For example, STD [578] uses sphere anchors for proposal generation, which has higher recall. In the instance segmentation method, the similar information is further subdivided. Usually, use proposal-based methods, first recognize 3D objects, and then instance mask prediction, for example, [594] uses the self-attention block to learn the feature representation of the point cloud bird's-eye view and then obtains the instance label according to the predicted horizontal center and height restrictions. Multi-step processing of proposal-based methods is usually time-consuming and computationally expensive. Therefore, proposal-free methods directly perform further processing on the basis of semantic segmentation. For example, Jiang et al. [239] proposed the PointGroup network, which is composed of a semantic segmentation branch and an offset prediction branch and further uses dual clustering algorithm and scoring network to obtain better grouping results.

5.1.4.2   Data Description and Recommendation

The publicly available datasets have played a key role in pushing forward the development in many computed vision-based image analysis tasks by providing problem-specific examples and the corresponding ground truth [101, 307]. In the context of lane detection for autonomous vehicles, TuSimple [526] and BDD100k [586] are the two commonly used benchmarks for the lane detection research community to evaluate diverse algorithms on the same data and contributed to closing the gap between the laboratory testing environment and diverse real-world problems in the lane detection field. A series of research papers and experiments are conducted based on these three public datasets.

In traffic scenes, the methods of image analysis using semantic markup methods include pixel-level semantic detection and instance-level semantic detection. The former aims to identify the type of each pixel, while the latter is to identify each instance.

Strengthening the semantic understanding of the surrounding environment is an important research direction, and there have been many influential datasets. Most datasets can be used for pixel-level semantic analysis. KITTI [163] is the most classic dataset, which provides semantic analysis in both line and sense, while the ApolloScape dataset [225] is specifically designed for semantic analysis and provides Scene Parsing 3D Car Instance and Lane Segmentation. TuSample [526] is designed for pixel-level line detection, with 7K more frames. BDD100K supports the semantic research of feasible domains and provides the richest semantic segmentation scene. Cityscapes [94] provides pixel-level research with two precisions, although it supports instance-level, but the semantic label is limited to people and cars. Mapillary [379] collects global multi-weather and multi-season daytime data around the semantic analysis of street scenes. IDD [535] dataset uses 4-level semantic tagging to deal with the challenge of semantic understanding of unstructured scenes. Semantic labels based on original images have always been marked by deviations due to image quality and other issues. SYNTHA [447] uses virtual technology to generate semantic data of virtual images. This brand-new semantic tag data provides high-quality semantic tags.

Among the datasets currently counted, only KITTI benchmark supports semantic segmentation of LiDAR point cloud data, which is also the dataset used by many researchers to carry out related research. This subset is collected using Velodyne HDL-64E, a total of 22 sequences, 80 GB.

5.1.5   Stereo

5.1.5.1   Algorithms

3D environment perception is an important research aspect. The acquisition methods are generally LiDAR-based and camera-based. Although the former is more effective, the latter has a great price advantage and is an important research direction. There are two categories around camera-based 3D scene algorithms, one is stereo- and monocular-based depth estimation, and the other is image-based 3D [547].

Stereo- and Monocular-Based Depth Estimation For stereo- and monocular-based depth estimation, based on the research of monocular depth estimation algorithms for many years, the accuracy has been significantly improved; for example, the algorithm DORN has very low errors in pixel depth prediction. In terms of stereo, the latest is MSMD-Net [482], which constructs a four-dimensional combination and a three-dimensional twist from multi-scale and multi-dimension and combines the residual network to improve the disparity estimation performance, making the outlier rate less than 1.5%.

Image-Based 3D Image-based 3D is that most of the existing algorithms are generally based on 2D object detection and use additional constraints to achieve 3D [365, 566]. The detection accuracy of previous algorithms is not improved, [547] believes that the low detection accuracy of the stereo algorithm in the past is not in the accuracy of the data but in the representation of the data, converted image-based depth maps into pseudo-radar representations, and made breakthrough progress, and in the KITTI dataset, the accuracy of object recognition within 30m was increased from 22 to 74%.

5.1.5.2   Data Description and Recommendation

To carry out research on stereo cameras, the first step is to label the data. In terms of stereo datasets in the field of autonomous driving, KITTI provides stereo benchmark consisting of 194 training image pairs and 195 test image pairs, and the dataset size reaches 19G. The ApolloScape dataset also provides labels, which is collected from consists 5165 image pairs and corresponding disparity maps, where 4156 image pairs are used for training, and 1009 image pairs are used for testing.

5.1.6   Localization

5.1.6.1   Algorithms

The positioning of vehicles in unknown environments is a crucial aspect of autonomous driving research and is one of the important foundations for vehicle safety research such as collision avoidance and attitude adjustment. How to reduce the uncertainty of positioning and improve the mapping ability that expands with the environment is the main challenge.

Camera-Based Algorithms Simultaneous localization and mapping (SLAM) algorithm is used to solve the positioning problem. Its key idea is Bayes' theorem

[367], which is to calculate the probability estimate of the next location based on the current state and input. The existing SLAM algorithms can be divided into two categories: filter-based and optimization-based [504].

Filter-Based SLAM Filter-based SLAM is based on the Bayesian filter with a two-step process iteration [165]. The first step is to use the evolution model and control input to estimate the vehicle's posture and position on the map. The second step is to compare the current sensor data measurement with the map to correct the prediction deviation that may exist in the first step. These steps will be repeated every time a new measurement is updated. Common algorithms are EKF [568] and FastSLAM [361].

Extended Kalman filter has the advantages of linearization to deal with nonlinear models and at the same time good closed loop and a lot of research work. These advantages make EKF the most important method to solve SLAM problems. But the algorithm will have problems when processing large maps [124]. Unlike EKF, FastSLAM does not require the Gaussian pose distribution, so it has the advantage of being able to accommodate any distribution, and the use of Rao–Blackwellized filter can achieve higher calculation speed; in addition, FastSLAM is a closed loop approach in solving SLAM problems. However, FastSALM has the problem of sampling data degradation and particle loss in the sampling process [595].

Optimization-Based SLAM Optimization-based SLAM also adopts a two-step iterative approach. The difference is that the first step detects the constraints of the problem, including the current position of the vehicle and sensor information. The second step is to calculate and optimize the posture of the vehicle and the previous posture on the map based on the constraints of the problem, so as to obtain the unity of the vehicle and the map. The important algorithm is Graph SLAM [185].

Graph SLAM contains the matrix structure of vehicle posture and landmarks in the map, so that the entire trajectory can be visualized; thanks to the Graph minimum optimal algorithm, the next position of the vehicle can be estimated with higher accuracy and better consistency. The disadvantage of Graph SLAM is that it requires a lot of computation.

Autonomous vehicles realize precise positioning, which requires information fusion of LiDAR, high-precision maps, GPS/IMU, etc. The method can usually be divided into two steps: laser radar and high-precision map registration and multi-sensor fusion. In the first step, the registration of LiDAR and high-precision map usually uses geometric matching and laser reflectivity matching methods. The second step is to achieve high accuracy and recall; multi-sensor fusion uses Bayesian filters for multi-mode fusion.

5.1.6.2   Data Description and Recommendation

In the field of positioning, KITTI still has an important place. It provides 22 stereo sequences (11 sequences (00–10) with ground truth trajectories for training and 11 sequences (11–21) without ground truth for evaluation.) There are two types of sub-data for SLAM research. The grayscale version data is 22 GB and the color version reached 65 GB. The ApolloScape dataset contains traffic, weather, and lighting conditions in a scene.

The specific tasks of L4 and L5 autonomous driving are very detailed and complicated. On the one hand, researchers divide the tasks to solve one by one, and on the other hand, they use an end-to-end approach. End-to-end model training requires massive sensor data and vehicle control information, and the training process is lengthy. The end-to-end approach can achieve automatic driving in common scenarios in one step, but this approach has an algorithmic black box and cannot better solve the problem of small probability events. In addition, the driver is an important research object, and how to link the driver's reaction with the information of the scene around the car is also a very meaningful attempt. The emergence of datasets for multi-vehicle driving provides a new research direction for the development of multi-vehicle collaboration.

5.1.7   Advance Intelligent Services

5.1.7.1   Algorithms

To carry out autonomous driving research, one research idea is to modularize tasks and subdivide tasks one by one to study solutions, such as object detection module, positioning module, planning module, vehicle control module, etc. Task segmentation is bound to increase the amount of calculation and inefficiency. And the task with the undertaking relationship will magnify the error step by step. These problems bring challenges to the research thinking of task refinement. Another research idea is to adopt an end-to-end approach. This research method avoids the intermediate links of data gradual processing and analysis and directly loads raw data such as camera/LiDAR/Radar and real-time vehicle control information (steering, acceleration, brake) into the end-to-end model for training. This method can avoid details and accelerate research and development. It is an increasingly popular research direction for autonomous driving.

5.1.7.2   Data Description and Recommendation

In the end-to-end task of autonomous driving, the Comma.ai [458] dataset collected 7.25 h of video and behavior information on the highway, which includes the speed, acceleration, steering angle, GPS coordinates, and gyroscope angles of vehicle. Simultaneously, 10 video clips of variable size are recorded at 20 Hz. The DBNet dataset provides the first end-to-end dataset that combines 2D and 3D perception with camera + LiDAR + behavioral information for 3D scenes, which has more than 1 TB data.

5.1.8 Summary

Datasets are essential for autonomous driving research. Own to the upgrading of devices (camera, LiDAR, radar, etc.) and the advancement of deep learning technology, especially computer vision research, it has greatly promoted the research of autonomous driving. With the deepening of L4 and L5 autonomous driving research, research tasks are gradually refined to improve the adaptability of autonomous driving, and more and more task-specific datasets are proposed. These datasets are different in terms of sensors, covered scenarios, label types, etc. The correspondence between autonomous driving datasets and autonomous driving tasks has not been well answered. To address this issue, we analyze 50 available publicly available dataset and organize them into several main autonomous driving tasks, and we provided the algorithm for the related tasks and the dataset choosing suggestions. Among them, please refer to Table 5.2 for the mapping relationship between tasks and datasets. We provide a reference for the selection of datasets for autonomous driving researchers to carry on new research areas, as well as a reference for researchers to switch to more suitable datasets.

5.2 CAVBench: A Benchmark Suite for Connected and Autonomous Vehicles

5.2.1 Introduction

With the rapid development of computer vision, deep learning, mobile communication, and sensor technology, the functions of vehicles are no longer limited to driving and transportation but have gradually become an intelligent, connected, and autonomous system. We refer to these advanced vehicles as connected and autonomous vehicles (CAVs). The evolution of vehicles has given rise to numerous new application scenarios, such as Advanced Driver Assistance Systems (ADAS) or Autonomous Driving (AD) [163, 454], Internet of Vehicles (IoV) [166], Intelligent Transportation Systems (ITS) [111], etc. Especially, for ADAS/AD scenarios, many industry leaders have recently published their own autonomous driving systems, such as Google Waymo [552], Tesla Autopilot [512], and Baidu Apollo [31].

Under these scenarios, the CAVs system becomes a typical edge computing system [483, 484]. The CAVs computing system collects sensor data via the CAN bus and feeds the data to on-vehicle applications. In addition, the CAVs system is not isolated in the network, so the CAVs will communicate with cloud servers

[318], Roadside Unit (RSU), and other CAVs to perform some computing tasks collaboratively. Much research focusing on the edge computing on CAVs from the application aspect has emerged [247, 286, 421]. There have also been some pioneer studies about exploring the computing architecture and systems for CAVs. NVIDIA®DRIVE™PX2 is an AI platform for autonomous driving that equips two discrete GPUs [389]. Liu et al. proposed their computing architecture for CAVs, which fully used hybrid heterogeneous hardware (GPUs, FPGAs, and ASICs) [319]. Unlike other computing scenarios, edge computing is still a booming computing domain. However, to date, a complete, dedicated benchmark suite to evaluate the edge computing platforms designed for CAVs is missing, both in academic and in industrial fields. This makes it difficult for developers to quantify the performance of platforms running different on-vehicle applications, as well as to systematically optimize the computing architecture on CAVs or on-vehicle applications. To address these challenges, we propose CAVBench, the first benchmark suite for edge computing systems on CAVs.

CAVBench is a benchmark suite for evaluating CAVs computing system performance. It takes six diverse real-world on-vehicle applications as evaluation workloads, covering four applications scenarios summarized in OpenVDAP: autonomous driving, real-time diagnostics, in-vehicle infotainment, and third-party applications [600]. The six applications that we have chosen are simultaneous localization and mapping (SLAM), object detection, object tracking, battery diagnostics, speech recognition, and edge video analysis. We collect four real-world datasets for CAVBench as the standard input to the six applications, which include three types of data: image, audio, and text. CAVBench also has two categories of output metrics. One is application perspective metric, which includes the execution time breakdown for each application, helping developers find the performance bottleneck in the application side. Another is system perspective metric, which we called the quality of service—resource utilization curve (QoS-RU curve). The QoS-RU curve can be used to calculate the Matching Factor (MF) between the application and the computing platform on CAVs. The QoS-RU curve can be considered as a quantitative performance index of the computing platform that helps researchers and developers optimize on-vehicle applications and CAVs computing architecture. Furthermore, we analyze the characteristics of the applications in CAVBench. We observe the application information and conclude three basic features of the applications on CAVs. First, the CAVs application types are diverse, and the real-time applications are dominated in CAVs scenarios. Second, the input data of CAVs applications is mostly unstructured. Third, deep learning applications in CAVs scenarios prefer end-to-end models. Then, we comprehensively characterize the six applications in CAVBench via several experiments. On a typical state-of-the-practice edge computing platform, we have the following conclusions:

- The operation intensity of applications in CAVBench is polarized. The deep learning applications have higher floating point operation intensity because the neural networks are their main workloads, which includes plenty of floating point multiplications and additions. As for computer vision applications, the algorithms rely more on mathematical models, which contains lower floating point operation intensity. Hence, the computing platform in CAVs should contain heterogeneous hardware to process these different applications.

- Similar to traditional computing scenarios, the applications in CAVBench need high memory bandwidth. That will cause competition for memory bandwidth when multiple real-time applications are running concurrently in real environments.
- On average, the CAVBench has a lower cache and TLB miss rate, which means the applications in CAVs scenarios have good data/instruction locality, but for specific workloads, some applications have one or two higher miss rates. Thus, the computing systems targeting these applications should value the optimization of the cache architecture and the data/instruction locality to improve the application performance.

Finally, we use the CAVBench to evaluate a typical edge computing platform and present the quantitative and qualitative analysis of the evaluation results of this platform.

5.2.2   Related Work

CAVBench is designed for evaluating the performance of the computing architecture and system of connected and autonomous vehicles. In this section, we summarize the related work from two aspects: the CAVs computing architecture and system and the benchmark suite related to CAVs.

5.2.2.1   Architecture and System for CAVs

Junior [362] was the first work to introduce a complete system of self-driving vehicles, which included type and location of sensors, as well as software architecture design [292]. Junior presented dedicated and comprehensive information about applications and a software flow diagram for autonomous driving. However, Junior provided less information about the computing system on its self-driving vehicles.

Liu et al. proposed a computer architecture for autonomous vehicles which fully used hybrid heterogeneous hardware [319]. In this work, the applications for autonomous driving were divided into three stages: sensing, perception, and decision-making. They compared the performance of different hardware running basic autonomous driving tasks and concluded some rules to perform different tasks for dedicated heterogeneous hardware. Lin et al. explored the architectural constraints and acceleration of autonomous driving system in [305]. They presented a detailed comparison of accelerating related algorithms using heterogeneous platforms including GPUs, FPGAs, and ASICs. The evaluation metrics included running latency and power, which will help developers build an end-to-end autonomous driving system that meets all design constraints. OpenVDAP [600] proposed a vehicle computing unit, which contained a task scheduling framework and heterogeneous computing platform. The framework scheduled the tasks to specific acceleration hardware, according to task computing characteristics and hardware utilization. In the industrial field, there are several state-of-the-practice computing platforms designed for CAVs, such as NVIDIA®DRIVE™PX2 [389] and Xilinx®Zynq®UltraScale+™ZCU106 [232].

These projects can be regarded as pioneering research in exploring the computing architecture and systems for connected and autonomous vehicles from different aspects. However, the evaluation method of these systems lacks uniform standards; all the research groups chose application type and implementation from their perspectives. Hence, it is challenging to evaluate and compare these systems fairly and comprehensively.

5.2.2.2 Benchmark Suite Related to CAVs

There are many classic benchmark suites in the traditional computing field, such as BigDataBench [544] for big data computing, Parsec [47] for parallel computing, HPCC [330] for high-performance computing, etc. However, for the computing scenario in CAVs, the benchmark research work is still at the beginning stage and can be divided into two categories according to their contents: datasets and workloads.

KITTI [162, 163] was the first benchmark dataset related to autonomous driving. It comprised rich stereo image data and 2D/3D object annotated data. According to different data types, it also provided a dedicated method to generate the ground truth and calculate the evaluation metrics. KITTI was built for evaluating the performance of algorithms in the autonomous driving scenario, including but not limited to optical flow estimation, visual odometry, and object detection. There are some customized benchmark datasets for each algorithm, such as TUM RGB-D [497] for RGB-D SLAM, PASCAL3D [567] for 3D object detection, and the MOTChallenge benchmark [284, 356] for multi-target tracking. These kinds of benchmark suites will help us choose the implementations and datasets of CAVBench.

Another class of related benchmark suites used a set of computer vision kernels and applications to benchmark novel hardware architectures. SD-VBS [537] and MEVBench [89] both are system performance benchmark suites based on computer vision workloads in diversified fields. SD-VBS assembled 9 high-level vision applications and decomposed them into 28 common computer vision kernels. It also provided single-threaded C and MATLAB implementations of these kernels. MEVBench focused on a set of workloads related with visual recognition applications including feature extraction, feature classification, object detection, tracking, etc. MEVBench provided single- and multi-threaded C++ implementations for some of the vision kernels. However, these two benchmarks are prior works in the field, so they are not targeted toward heterogeneous platforms such as GPUs. SLAMBench

[375] concentrated on using a complete RGB-D SLAM application to evaluate novel heterogeneous hardware. It chose KinectFusion [383] as the implementation and provided C++, OpenMP, OpenCL, and CUDA versions of key function kernels for different platforms. The RGB-D cameras are more suitable for indoor environments, and the workload type in SLAMBench is single. These efforts are a step in the right direction, but we still need a comprehensive benchmark which contains diverse workloads that cover varied application scenarios of CAVs to evaluate the CAVs system as we mentioned above.

5.2.3   Benchmark Design

The objective of developing CAVBench is to help developers determine if a given computing platform is competent for all CAVs scenarios. This section presents the methodology, overview, and components of our CAVBench.

5.2.3.1   Methodology and Overview

Combined with the survey and analysis of the related works on CAVs architectures and systems and existing benchmark suites, we present our methodology for designing CAVBench as shown in Fig. 5.1. The computing and application scenarios of connected and autonomous vehicles are much different from the traditional domain. First, we investigate the typical application scenarios of CAVs. It is well known that Advanced Driver-Assistant Systems (ADAS) and Autonomous Driving (AD) have already become a dominant application scenario of CAVs [41, 163, 292, 531]. In addition to ADAS/AD, OpenVDAP summarizes three other scenarios that are Real-time Diagnostics (RD), In-Vehicle Infotainment (IVI), and Third-Party Application (TApp) [600]. Thus, we focus on the exemplary and key applications in each dominant scenario.

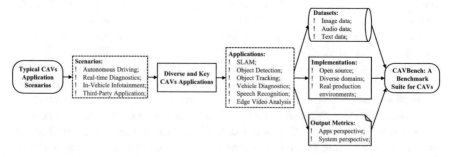

Fig. 5.1 CAVBench methodology

The tasks in the ADAS/AD scenario can be divided into three stages according to their functions: sensing, perception, and decision-making [319]. Sensing tasks manage and calibrate the various sensors around the CAVs and provide reliable sensing data to upper-level tasks. Perception tasks take the sensing data as the input and output the surrounding information to the decision-making tasks, which in turn generate a safe and efficient action plan in real-time. It can be seen that perception is an important connecting link between sensing and decision-making. The three main perception tasks are simultaneous localization and mapping (SLAM), object detection, and object tracking, which are all visual-based applications. Many studies take them as the vital parts in the autonomous driving pipeline [163, 254, 305]. Hence, we chose these three applications in the ADAS/AD scenario.

Vehicle system fault diagnosis and prognosis are important for keeping vehicles stable and safe [605]. With the development and widespread use of electric and hybrids vehicles, the health monitoring and diagnostics of Li-ion batteries in these vehicles have received increasingly more attention [597]. It is extremely important to monitor and predict the battery status in a real-time fashion, including multiple parameters of each battery cell, e.g., voltage, current, temperature, and so on. Thus, battery diagnostics will be the application chosen in the RD scenario.

The In-Vehicle Infotainment (IVI) scenario includes a wide range of applications that provide audio or video entertainment. Compared with manual interaction, the speech-based interaction method reduces the distraction of drivers and ensures driving safety [337], so motor companies have increasingly begun to develop their own IVI systems with speech recognition such as Ford® SYNC® [145]. We chose speech recognition applications for the IVI scenario.

There are some preliminary projects for the third-party application scenario. PreDriveID [249] is a driver identification application based on in-vehicle data. It can enhance vehicle safety by detecting whether the driver is registered or not through by analyzing how a driver operates a vehicle. A3 [601] is an edge video analysis application which uses a vehicle on-board camera to recognize targeted vehicles to enhance the AMBER alert system. It is easier to acquire the data from an on-board camera than the vehicle bus data, and edge video analysis could well be a killer application for edge computing [12], so we chose this kind of application for the TApp scenario.

After selecting the six applications, we pay attention to the implementation, datasets, and output metrics for the applications. The implementation of each application should be state-of-the-art and representative, ensuring that it can be deployed in a real production environment. We provide real-world datasets for each application which are open source or collected by ourselves to let the applications have a standard input. To give the user a complete understanding of the benchmark results, the output metrics contain two categories: application perspective metric and system perspective metric. These three parts (implementation, datasets, and output metrics) form the CAVBench and will be introduced in detail in the next three subsections, respectively.

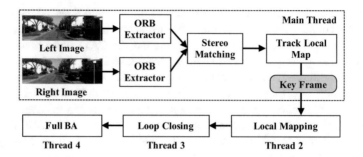

Fig. 5.2 Overview of the ORB-SLAM2 pipeline

5.2.3.2 Implementation

The implementation we chose in CAVBench is shown in Table 5.3. The reasons for choosing these implementations are presented as follows.

SLAM The simultaneous localization and mapping (SLAM) technique helps CAVs with real-time building a map of an unknown environment and localizing themselves in the map. ORB-SLAM2 provides a stereo SLAM method, which has been ranked as the top in KITTI benchmark datasets according to the accuracy and runtime. ORB-SLAM2 is more suitable for large-scale environments than monocular [369] and RGB-D SLAM [128], so we chose it as the SLAM implementation. Figure 5.2 shows the ORB-SLAM2 pipeline. The stereo image stream is fed into the ORB extractor to detect feature points and generate descriptions of the extracted feature points. Then, the main thread attempts to match the current descriptions with the prior map point to localize and generate new keyframes. The local mapping thread manages keyframes to create new map point. The loop closing thread tries to detect and close trajectory loops via the last keyframe processed by the local mapping and creates the fourth thread to optimize a global map by the full bundle adjustment (BA).

Object Detection The visual object category recognition and detection have always been a challenging problem in the last decade [133]. In recent years, the series of algorithms based on convolutional neural networks (CNNs) have become one of the mainstream techniques in the field of object detection [65, 441], especially in the autonomous driving scenario [78]. Single Shot Multibox Detector (SSD) [320] and You Only Look Once (YOLO) [434] are kinds of end-to-end CNN model. Compared with the R-CNN series [172], they do not hypothesize bounding boxes or resample pixels or features for these hypotheses, which improves the speed for detection and is as accurate as the R-CNN series. The network structure of SDD is shown in Fig. 5.3. SSD uses multiple feature maps from the different stages of the network to perform detection at multiple scales, which is more accurate than the detection by one full connection layer in YOLO. In a word, SSD has higher

Table 5.3 Overview of implementation in CAVBench

Scenario	Application	App type	Implementation	Main workloads	Data type	Data source
ADAS/AD	SLAM	Real-time	ORB-SLAM2 [371]	ORB extractor and BA	Unstructured	Image (Stereo)
ADAS/AD	Object detection	Real-time	SSD [320]	CNNs	Unstructured	Image (Monocular)
ADAS/AD	Object tracking	Real-time	CIWT [398]	EKF and CRF model	Unstructured	Image (Stereo)
RD	Battery diagnostics	Offline	EVBattery	LSTM networks	Semi-structured	Text
IVI	Speech recognition	Interactive	DeepSpeech [196]	RNNs	Unstructured	Audio
TApp	Edge video analysis	Interactive	OpenALPR [394]	LBP feature detector	Unstructured	Image (Monocular)

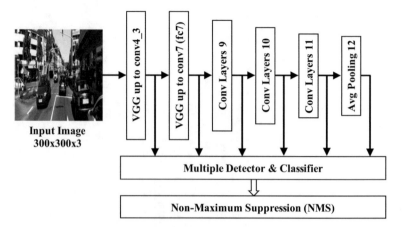

Fig. 5.3 Overview of the single shot multibox detector network structure

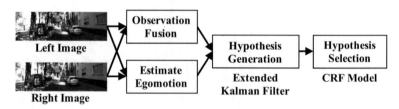

Fig. 5.4 Overview of the combined image- and world-space tracking pipeline

accuracy and processing speed than other models; hence, we chose SSD as the implementation for object detection.

Object Tracking The main goal of object tracking is to ensure that the vehicle does not collide with a moving object, whether a vehicle or a pedestrian crossing the road. We chose Combined Image- and World-Space Tracking (CIWT) [398] as the implementation for object tracking, which is an online multiple object tracking application dedicated to the autonomous driving scenario. In KITTI results, CIWT is not the most accurate, but it costs fewer computation resources and less processing time than some algorithms ranked ahead of it. It is more practical in the real-world environment than some offline [161] or single target tracking algorithms [202]. Figure 5.4 shows the overview of the CIWT pipeline. CIWT uses a stereo image stream to fuse the observation and estimate the egomotion. The observation fusion includes 2D detection and 3D proposals. The tracking process uses these results to generate tacking hypotheses through the extended Kalman filter (EKF) and uses the conditional random field (CRF) model to select a high score tacking hypothesis.

Battery Diagnostics The devices monitor and log data is a kind of temporal data. Recently, some works leverage Long-Short Term Memory (LSTM) networks

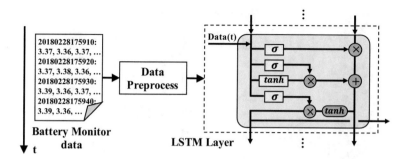

Fig. 5.5 Overview of the EVBattery diagnostics

to perform failure prediction according to log data for hard drives [118]. LSTM belongs to recurrent neural networks (RNNs), but it has better performance on the long-term prediction task than RNNs. We use the similar method to process EV battery data. We call our implementation as EVBattery Diagnostics and the process is shown in Fig. 5.5.

Speech Recognition Voice service generally consists of two steps: speech to text and text to intent, with the former being the process of speech recognition. Deep-Speech [196] is an end-to-end speech recognition algorithm based on RNNs. The deep learning method supersedes traditional processing stages in speech recognition systems, such as those that have been hand-engineered. Figure 5.6 shows the DeepSpeech network structure. The first three and the fifth layers in DeepSpeech have the basic full connection structure. The fourth layer is a bidirectional recurrent layer which is used to characterize the temporal correlation of the voice data. The evaluation results show that DeepSpeech has less latency and error rates than traditional methods based on the Hidden Markov Model (HMM) [336], especially in noisy environments. Therefore, DeepSpeech is a suitable implementation for speech recognition.

Edge Video Analysis As we mentioned above, AMBER Alert Assistant (A3) is a typical edge video analysis application that takes OpenALPR [394] as the core workload to detect target vehicles in the video. The OpenALRP pipeline is shown in Fig. 5.7. OpenALRP is a classic implementation of automatic license plate recognition, including several typical computer vision modules: LBP detector, image deskew, and ORC. Therefore, we chose OpenALPR as the implementation of edge video analysis.

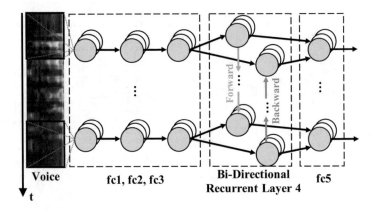

Fig. 5.6 Overview of the DeepSpeech network structure

Fig. 5.7 Overview of the OpenALPR pipeline

Table 5.4 The summary of datasets in CAVBench

Application	Datasets	Data size and type
SLAM	KITTI VO/SLAM Datasets [162]	21 sequences stereo grayscale image
Object detection	KITTI object detection datasets [162]	7518 monocular color image
Object tracking	KITTI object tracking datasets [162]	28 sequences stereo color image
Battery diagnostics	EV battery monitor data	30 days battery monitor data, 160,000 rows
Speech recognition	Mozilla corpus [366]	3995 valid human voice data
Edge video analysis	Images of vehicles with license plates	1000 monocular color image

5.2.3.3 Datasets

To guarantee that the evaluation results are similar to running in the real environment, we need to choose a real-world, not a synthetic, dataset for each application. Table 5.4 shows the basic information of the datasets we have chosen. It must be noted that we collected the datasets of Battery Diagnostics and Edge Video Analysis by ourselves because of the lack of relative open-source datasets or some datasets did not meet our real-world requirements.

KITTI Datasets [162] KITTI provides rich, open-source, and real-world image data for different autonomous driving applications. The image characteristics varies, because each dataset focuses on one specific application. Each dataset contains various traffic scenes which can evaluate the application performance comprehensively.

EV Battery Monitor Data There are few datasets that provide vehicle battery monitoring data. We collect the battery monitoring data of an electric vehicle in the real environment for 1 month. Each record of the data contains 60 items, such as voltage and temperature.

Mozilla Corpus [366] The Mozilla corpus provides 3995 valid common voice files for testing. Each file is a record in which a person reads a common sentence, and records are collected by numerous people reading different sentences. It must be noted that the Mozilla corpus still has some limitations. It was collected in a daily environment, so it may not contain words that are used in the vehicular setting and may not have enough background noise that is likely to be very common in a vehicular environment.

Images of Vehicles with License Plates The images in KITTI datasets could not meet the resolution requirement of performing license plate recognition, and some license plate datasets are not collected by a vehicle on-board camera. Thus, we use the Leopard® LI-USB30-AR023ZWDRB video camera [290] with 6 and 25 mm lens, which was suggested by the Apollo project [31] to collect image data in real traffic scenes. Each image we provided contains at least one vehicle with its license plate.

5.2.3.4 Output Metrics

The output metrics show quantitatively whether the given hardware platform can be used for CAVs scenarios or not. In CAVBench, the output metrics contain two parts: application perspective metric and system perspective metric.

Application Perspective Metric Like some traditional benchmark suites, the application perspective metric shows the running time of each application. For computer vision applications (ORB-SALM2, CIWT, and OpenALPR), we output the average latency for each module in the applications, and we provide the average and tail latency for deep learning applications (SSD, EVBattery, and DeepSpeech). This metric helps developers optimize the platform in terms of applications.

System Perspective Metric For the system perspective metric, we call it the quality of service—resource utilization curve (QoS-RU curve). We evaluate the QoS of each application under different system resource allocations and draw the QoS-RU curve for each system resource (CPU utilization, memory footprint, memory bandwidth, etc.). Figure 5.8 shows an example of the QoS-RU curve. We use the area under the curve of each system resource to calculate the Matching Factor (MF) between the application and the platform, indicating whether the platform

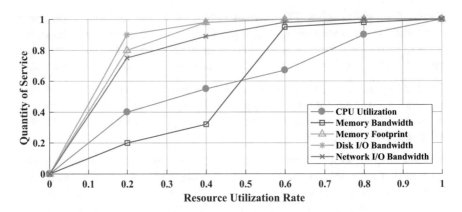

Fig. 5.8 An example of QoS-RU curve

is suitable for the CAVs application. Following is our approach to calculate the Matching Factor:

We denote the area under the curve of each system resource as A_i, and we take the weighted average of each area as the Matching Factor M, as Eq. 5.1 shows.

$$M = \sum_{i=1}^{n} (w_i \cdot A_i) \tag{5.1}$$

The weight for each resource w_i can be calculated by Eq. 5.2, in which n is the number of system resources we considered.

$$w_i = (1 - A_i) / \sum_{i=1}^{n} (1 - A_i) \tag{5.2}$$

We notice that the w_i is the normalized $1 - A_i$. If A_i is large, the resource i is relatively sufficient for the application. Similarly, if A_i is small, the resource i has the potential to be the bottleneck of the platform. Thus, A_i and its weight have opposite values, which is why we chose the normalized $1 - A_i$ as the weight.

5.2.4 Benchmark Characterization

In this section, we present the detailed description of the experiments of our CAVBench workload characterization analysis.

Overview Before we did the characterization analysis experiments, we first observed the information of the six applications presented in Table 5.3 to conclude some basic features of the applications in CAVs scenarios. The observations are described as follows:

Real-Time Applications are Dominant   The application types in the CAVs computing scenarios are diverse, including real-time, offline, and interactive. This diversity corresponds to cloud computing and the big data computing domain [544]. In contrast to traditional computing fields, real-time applications are dominant in CAVs, and they all belong to the ADAS/AD scenario. Furthermore, the applications in other scenarios are offline and interactive. That explains why ADAS/AD applications always have the highest priority.

Unstructured Data Type   The input data type of CAVs applications is mostly unstructured data. As we mentioned above, CAV is a typical embedded and edge computing system, and it deploys various sensors to collect information from the real physical world and uses the information (data) to execute computation tasks. Therefore, the input data is generally unstructured, such as images and audios. Even for vehicle monitor data, it usually has little structural constraints, so we consider it as semi-structured data. As for cloud computing, there are still some classic workloads that use the structured data, for example, the relational query operation.

End-to-End Deep Learning Workloads   According to the classification of the main workloads in CAVBench, we find that the workloads in CAVs all belong to computer vision, machine learning, and the deep learning domain because the main functions of the applications in CAVs are detection, recognition, and prediction. In addition, due to the limitation of latency, deep learning workloads in CAVs choose end-to-end models, which means that with the exception of deep neural networks, there are no other processes between input and output; this improves the running speed while not decreasing the algorithm accuracy.

[Insights]   These three observations can be regarded as the basic characteristics of the applications in the CAVs computing scenario. We can conclude some insights regarding the CAVs computing system and applications design. First, real-time applications have the highest priority in real production environments, so the CAVs system should contain a task scheduling framework to ensure that the real-time applications can be allocated with enough computing resources. Second, pre-processing the unstructured data consumes more time, so a hardware accelerator aimed at transforming the unstructured data to structured will be a benefit to the performance of the whole CAVs system. Third, the end-to-end deep learning algorithm reduces the number and frequency of data movements (main memory to GPU memory), decreasing the processing latency. Thus, this kind of algorithm will be more suitable for CAVs applications.

5.2.4.1   Experiments Configurations

To obtain insights regarding application characteristics in CAVBench, we ran the six applications in a typical edge device and use the Linux profiling tool Perf to

Table 5.5 Edge computing
platform configurations

Platform	Intel FRD
CPU	Intel Xeon E3-1275 v5
Number of sockets	1
Core(s) per socket	4
Thread(s) per core	2
Architecture	X86_64
L1d cache	4×32 KB
L1i cache	4×32 KB
L2 cache	4×256 KB
L3 cache	8 MB
Memory	32 GB SODIMM 3122 MHz
Disk	256 GB PCIe SSD

collect the behaviors of the applications at the architecture level. We chose Intel®
Fog Reference Design (FRD) as the experiment platform, which has one Xeon®
E3-1275 v5 processor equipped with 32 GB DDR4 memory and 256 GB PCIe
SSD. The processor has four physical cores, and hyper-threading is enabled. Other
detailed information of the platform is shown in Table 5.5. The operating system
is Ubuntu 16.04 with Linux kernel 4.13.0. The deep learning applications are built
on TensorFlow 1.5.0, and some visual modules are implemented on OpenCV 3.3.1.
To acquire the pure and original characteristics of the applications, the platform
is not equipped with heterogeneous devices. Each application executes for 500 s,
sequentially processing different data.

5.2.4.2   Operation Intensity

First, we analyzed the operation intensity of CAVBench via the instruction breakdown of each application in CAVBench. As shown in Fig. 5.9, the distribution of the
instructions is diverse, even polarized. The difference is mainly due to floating point
instruction (FL). The average proportion of floating point instruction in CAVBench
is 20.77%, and the average ratio of integer instructions (Int) to FL is 24.44. However,
for each application, the minimum and the maximum FL proportions are 0.38%
(CIWT) and 48.89% (SSD), and the minimum and the maximum Int/FL ratios are
0.63 (SSD) and 79.59 (OpenALPR). In addition, the average FL proportion for
BigDataBench, HPCC, and Parsec is 2.12%, 24.11%, and 18.25%, respectively.
The instruction distribution in CAVBench is similar to HPCC, Parsec according
to the average values, but the distribution is also polarized when investigating
specific applications in CAVBench. In contrast, the instruction distribution of each
workload in the traditional benchmark is similar, such as the results presented in
BigDataBench [544].

In order to characterize the computation behaviors, we calculated the ratio of
computation to memory access for each application, which represents the integer

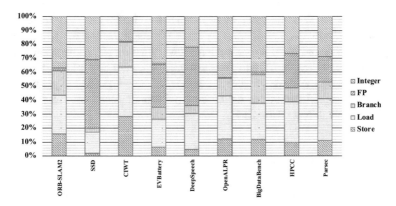

Fig. 5.9 Instruction breakdown

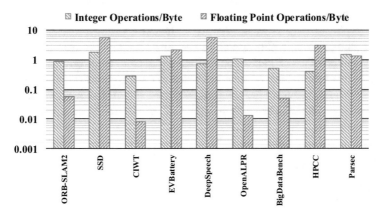

Fig. 5.10 Integer and floating point operation intensity

and floating point operation intensity. As shown in Fig. 5.10, the floating point operation (FLO) intensity of each application in CAVBench is still polarized, and the minimum and the maximum values are 0.0079 (CIWT) and 5.46 (SSD), respectively, which differ by about three orders of magnitude. As for integer operation (IntO) intensity, the minimum and the maximum values are 0.28 (CIWT) and 1.80 (SSD), and the IntO intensity for BigDataBench, HPCC and Parsec is 0.52, 0.43, and 1.50, respectively. Hence, the IntO intensity of CAVBench is almost in the same order of magnitude as those of the other benchmarks.

[Insights] According to operation intensity experiments, we can draw an important conclusion: the operation intensity of applications in the CAVs scenario is polarized. Deep learning applications, such as SSD and DeepSpeech, have higher floating point operation intensity, which is similar to the workloads in the high-performance computing domain. That is because the neural networks are the main workloads in deep learning applications, which includes a large number of matrix operations, causing plenty of floating point multiplications and additions. As for computer vision applications, the algorithms rely more on mathematical modeling, and the computation is not so high; thus, they have lower floating point operation intensity. This phenomenon explains why modern CAVs computing platforms leverage heterogeneous hardware to accelerate some tasks. The state-of-the-practice CPUs provide SSE, AVX instructions for floating point operations, but the performance does not yet match that of the GPUs.

5.2.4.3   Memory Behavior

Memory is a very important part of the computer system; the memory wall problem exists in many computing domains. Therefore, we further investigated the memory behaviors of CAVBench. Our experiments included three parts: memory bandwidth, memory footprint, and cache behavior, which we discuss below:

Due to the restriction of the hardware function in the processor, we cannot monitor memory bandwidth directly, so we use Perf tools to acquire the indirect memory bandwidth. The measuring method will lead to some errors, but it can still help us analyze the memory access behaviors of each application in CAVBench.

The real-time memory bandwidth of each application is shown in Fig. 5.11. The black line is the memory bandwidth upper limit of our experiment platform, whose

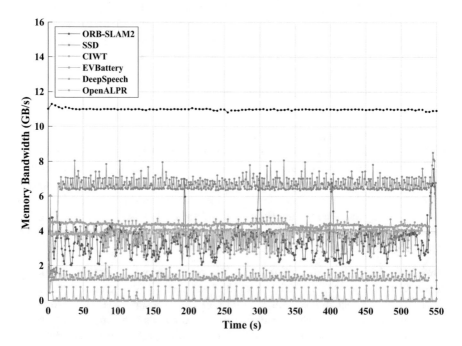

Fig. 5.11 Memory bandwidth behaviors

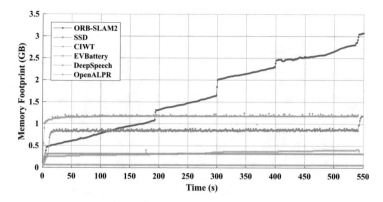

Fig. 5.12 Memory footprint behaviors

average value is 10.98 GB/s. The applications in CAVBench sequentially process the data in the same size and type, so they all have stable memory bandwidth, as shown in Fig. 5.11. The minimum bandwidth is 1.01 GB/s (EVBattery), which is 9.20% of the bandwidth upper limit, and the maximum value is 6.57 GB/s (SSD), which is 59.84% of the upper limit. According to this observation, we can see that the applications in the CAVs scenario require the high memory bandwidth, and the bandwidth may become the bottleneck of the performance in the real environment. When several CAVs applications run concurrently, they will compete for memory bandwidth resources, leading to higher latency for each application. We observe that the memory bandwidth of ORB-SLAM2 has four peak values during the running time, which reaches 7.21 GB/s on average. This is due to the loop closing module performing full BA when a loop is detected in the trajectory. This kind of memory access burst will cause interference in the performance of other applications, especially the tail latency.

Furthermore, we investigated the resident memory (memory footprint) behavior of CAVBench. The experiment results are shown in Fig. 5.12. Except for ORB-SLAM2, other applications have stable memory footprint. As we mentioned above, the total memory of the experiment platform is 32 GB. With the exception of ORB-SLAM2, the lowest memory footprint is 0.061 GB (OpenALPR), which is 0.19% of the total memory, and the highest memory footprint is 1.17 GB (DeepSpeech), which is 3.66% of the total memory. The reason for the continued increment of ORB-SLAM2 memory footprint is that the application continues to generate new map points and the point data stores in the memory. The four jumps of the memory footprint are also caused by the full BA, which corresponds to the observation in memory bandwidth experiment. The maximum memory footprint of ORB-SLAM2 is 3.07 GB (9.59%). We can conclude that the applications in the CAVs scenario consume less memory footprint, and the large capacity memory is available for the edge computing platform, so the CAVs application performance will not be constrained by the memory footprint.

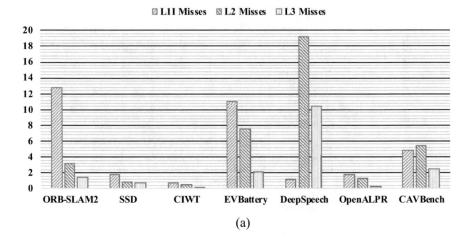

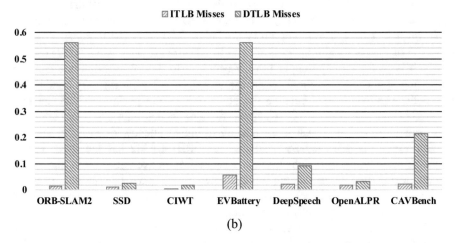

Fig. 5.13 (**a**) Cache behaviors and (**b**) TLB behaviors

Finally, we investigated the cache behaviors of CAVBench to see that whether the memory hierarchy architecture in the state-of-the-practice edge platform is proper for CAVs applications. The cache behavior and TLB behavior of each application are shown in Fig. 5.13. The CAVBench average L1i, L2, and L3 cache MPKI (Misses Per Kilo Instructions) are 4.86, 5.44, and 2.51, respectively, which are almost in the same order of magnitude as HPCC (0.41, 5.59, and 4.22) and Parsec (3.51, 7.25, and 3.37), respectively. The TLB behavior of CAVBench is also similar to HPCC and Parsec. According to this observation, we find that the applications in CAVBench have good data and instruction locality. These characteristics differ from the big data computing, which has a huge code size and deep software stack leading to higher MPKI.

Focusing on specific applications, ORB-SLAM2, EVBattery, and DeepSpeech have a higher MPKI than the other three. The ORB-SLAM2 has high L1i MPKI and DTLB MPKI. We infer that this phenomenon is still caused by the periodic loop detection operation and irregular full BA operation. The loop closing module queries the local map database periodically to detect the potential trajectory loop. The DLTB has less capacity to store all the page tables of the local map database, causing the high DTLB miss rate, and the irregular full BA operation interferes with the instruction locality, increasing the L1i cache miss rate. As for EVBattery and DeepSpeech, we infer that the RNN structure leads to a high cache miss rate. The convolution operations in CNNs make the data and instruction localized, which is why SSD does not have a high cache miss rate, but the RNNs do not have such convolution operations.

[Insights] According to the memory experiments, we can draw some important conclusions: first, applications in the CAVs scenario consume high memory bandwidth, which will be a performance bottleneck when multiple applications run concurrently in real environments. Second, the memory footprint of each application takes a very low proportion of the total memory in the state-of-the-practice edge computing platform. Third, on average, the applications in the CAVs scenario have good data and instruction locality. This characteristic is similar to the workload in high-performance computing and the parallel computing domain. As for specific applications, the SLAM and RNN model-based applications have a higher probability to increase the cache and TLB miss rate. Therefore, with the CAVs computing system paying more attention to these applications, we should focus on the optimization of the cache architecture and the application data/instruction locality.

5.2.5 Summary

CAVBench is the first benchmark suite for computing system and architectures designed for connected and autonomous vehicles targeting computational performance evaluation. We chose four typical and dominate application scenarios of CAVs and summarized six applications in these scenarios as the evaluation applications. After that, we collected state-of-the-art implementation and standard input datasets for each application and determined the output metrics of the CAVBench. We got three basic features from CAVBench. First, the CAVs application types are diverse, and the real-time applications are dominated in CAVs scenarios. Second, the input data is mostly unstructured. Third, the end-to-end deep learning algorithm is more preferable for CAVs computing system. Then, we ran a series of experiments to explore the characteristics of CAVBench and concluded three observations as follows. First, the operation intensity of the applications in CAVBench is polarized. Second, the applications in CAVBench all consume high memory bandwidth. Third, CAVBench has a lower cache miss rate on average, but for specific applications, the optimization of the cache architecture and data/instruction locality is still important. According to these features and characteristics, we presented some insights and suggestions about the CAVs computing system design or CAVs application implementation. Finally, we used the CAVBench to evaluate a typical edge computing platform and presented the quantitative and qualitative analysis of the benchmarking results.

We hope this work will be helpful to researchers and developers who target the computing system or architecture design of connected and autonomous vehicles. According to the insights proposed in this chapter, our future work will proceed from the following aspects. First, we will focus on providing the CUDA and OpenCL implementation of the CAVBench to support more heterogeneous platforms. Second, we will explore more methodologies to evaluate the computing system in the CAVs scenarios comprehensively, such as a benchmark dedicating system memory behaviors. Third, we will implement a full stack computing system for CAVs that will be competent for all CAVs applications.

Chapter 6
Autonomous Driving Simulators

6.1 Introduction

According to the annual Autonomous Mileage Report[1] published by the California Department of Motor Vehicles, Waymo has logged billions of miles in testing so far. As of 2019, the company's self-driving cars have driven 20 million miles on public roads in 25 cities and additionally 15 billion miles through computer simulations [555]. While the number of miles driven is important, it is the sophistication and diversity of miles accumulated that determines and shapes the maturity of the product [224]. Additionally, the testing through simulation plays a key role in supplementing and accelerating the real-world testing [555]. It allows one to test scenarios that are otherwise highly regulated on public roads because of various safety concerns [589]. It is reproducible and scalable and reduces the development cost.

There are many simulators available for testing the software for self-driving cars, which have their own pros and cons. Some of them include CarCraft and SurfelGAN used by Waymo [134, 553], Webviz and The Matrix used by Cruise, and DataViz used by Uber [518]. Most of these are proprietary tools; however, there are many open-source simulators available as well. In this chapter, we compare MATLAB/Simulink, CarSim, PreScan, Gazebo, CARLA, and LGSVL simulators with the objective of studying their performance in testing new functionalities such as perception, localization, vehicle control, and creation of dynamic 3D virtual environments. Our contribution is two-fold. We first identify a set of requirements that an ideal simulator for testing self-driving cars must have. Second, we compare the simulators mentioned above and make the following observations.

An ideal simulator is the one that is as close to reality as possible. However, this means it must be highly detailed in terms of 3D virtual environment and

---

[1] Autonomous Mileage Report: Disengagement Reports.

W. Shi, L. Liu, *Computing Systems for Autonomous Driving*,
https://doi.org/10.1007/978-3-030-81564-6_6

very precise with lower level vehicle calculations such as the physics of the car. So, we must find a trade-off between the realism of the 3D scene and the simplification of the vehicular dynamics [141]. CARLA and LGSVL meet this trade-off, making them the state-of-the-art simulators. Furthermore, Gazebo is also a popular robotic 3D simulator, but it is not very efficient in terms of the time involved in creating a 3D scene in the simulation environment. The simulators such as MATLAB/Simulink still play a key role because they offer detailed analysis of the results with their plotting tools. Similarly, CarSim is highly specialized at vehicle dynamic simulations as it is backed by more precise car models. The detail reasoning behind these observations is described in the chapter below.

6.2   Motivation

The complexity of automotive software and hardware is continuing to grow as we progress toward building self-driving cars. In addition to tradition testing such as proper vehicle dynamics, crash-worthiness, reliability, and functional safety, there is a need to test self-driving-related algorithms and software, such as deep learning and energy efficiency [310]. As an example, a Volvo vehicle built in 2020 has about 100 million lines of code according to their data [14]. This includes code for transmission control, cruise control, collision mitigation, connectivity, engine control and many other basic and advanced functionalities that come with the cars bought today. Similarly, the cars now have more advanced hardware, which includes plethora of sensors that ensure vehicles are able to perceive the world around them just like humans do [211]. Therefore, the complexity of the modern age vehicle is the result of both more advanced hardware and software needed to process the information retrieved from the environment and for decision-making capability.

Finally, in order to assure that the finished product complies with the design requirements, it must pass rigorous testing, which is composed of many layers. It ranges from lower level testing of Integrated Circuits (ICs) to higher level testing of vehicle behavior in general. The testing is accomplished by relying on both physical and simulation testing. As the complexity of vehicles continue to grow, so does the complexity, scale and the scope of testing that becomes necessary. Therefore, the simulators used for automotive testing are in a continuous state of evolving.

These simulators have evolved from merely simulating vehicle dynamics to also simulating more complex functionalities. Table 6.1 shows various levels of automation per the Society of Automotive Engineers (SAE) definitions [491], along with the evolving list of requirements for testing that are inherent in our path to full automation. It is important to note that Table 6.1 focuses on requirements that are essentially new to testing driver assisted features and autonomous behavior [472]. This includes things such as perception, localization and mapping, control algorithms, and path planning

Table 6.1 Testing requirements to meet SAE automation levels

SAE J3016 levels of driving automation		Testing requirements
Levels	Description	
Level 0	No automation: features are limited to, warnings and momentary assistance. Examples: LDW, blind spot warning	Simulation of: traffic flow, multiple road terrain type, radar and camera sensors
Level 1	Assisted: features provide steering OR brake/acceleration control. Examples: lane centering OR ACC	All of the above plus simulation of vehicle dynamics, ultrasonic sensors
Level 2	Partial automation: features provide steering AND brake/acceleration control. Examples: lane centering AND ACC at the same time	All of the above plus simulation of driver monitoring system. Human–machine interface
Level 3	Conditional automation: features can drive the vehicle when all of its conditions are met. Examples: traffic jam assist	All of the above plus simulation of traffic infrastructure, dynamic objects
Level 4	High automation: features can drive the vehicle under limited conditions. No driver intervention. Examples: local driverless taxis	All of the above plus simulation of different weather conditions, LiDAR, camera, radar sensors, mapping and localization
Level 5	Full automation: features can drive the vehicle in all conditions and everywhere. Examples: full autonomous vehicles everywhere	All of the above plus compliance with all the road, rules, V2X communication

These definitions are per the SAE J3016 Safety levels. More details can be found below: https://www.sae.org/standards/content/j3016_201806/
LDW = Lane Departure Warning
ACC = Adaptive Cruise Control

Thus, the simulators intended to be used for testing self-driving cars must have requirements that extend from simulating physical car models to various sensor models, path planning and control. Section 6.3 dives deeper into these requirements.

6.3 Methodology

The emphasis of this chapter is on testing the new and highly automated functionality that is unique to self-driving cars. This section identifies a set of criteria that can serve as a metric to identify which simulators are a best fit for the task at hand.

The approach we take to compile requirements for a simulator is as below. Firstly, we focus on the requirements driven by the functional architecture of self-driving cars [136] (Requirements 1–4). Secondly, we focus on the requirements that must be met in order to support the infrastructure to drive the simulated car (Requirements 5–7). Thirdly, we define the requirements that allow the use of simulator for secondary tasks such as data collection for further use (Requirement 8). Finally, we list generic requirements desired from any good automotive simulator (Requirement 9).

1. **Perception**: One of the vital components of self-driving cars is its ability to see and make sense (perceive) the world around itself. This is called perception. The vehicle perception is further composed of hardware, that is available in the form of wide variety of automotive grade sensors and software, that interprets data collected by various sensors to make it meaningful for further decisions. The sensors that are most prevalent in research and commercial self-driving cars today include camera, LiDAR, ultrasonic sensor, radar, Global Positioning System (GPS), and Inertial Measurement Unit (IMU) [534]. In order to test a perception system, the simulator must have realistic sensor models and/or be able to support an input data stream from the real sensors for further utilization. Once the data from these sensors is available within the simulation environment, researchers can then test their perception methods such as sensor fusion [266]. The simulated environment can also be used to guide sensor placement in a real vehicle for optimal perception.

2. **The Multi-view Geometry**   The simultaneous localization and mapping (SLAM) is one of the components of autonomous driving (AD) systems that focuses on constructing the map of unknown environments and tracking the location of the AD system inside the updated map. In order to support SLAM applications, the simulator should provide the intrinsic and extrinsic features of cameras. In other words, it should provide the camera calibration. According to this information, the SLAM algorithm can run the multi-view geometry and estimate the camera pose and localize AD system inside the global map.

3. **Path Planning**:The problem of path planning revolves around planning a path for a mobile agent so that it is able to move around autonomously without collision with its surroundings. The path planning problem for autonomous vehicles piggy backs on the research that has already been done in the field of mobile robots in the last decade. This problem is subdivided into local and global planning [176], where the global planner is typically generated based on a static map of the environment and the local planner is created incrementally based on the immediate surroundings of the mobile agent. In order to create these planners, various planning algorithms play a key role [229]. To implement such intelligent path planning algorithms like A*, D*, and RRT algorithms[176], the simulator should at least have a built-in function to build maps or have interfaces for importing maps from outside. In addition, the simulator should have interfaces for programming customized algorithms.

4. **Vehicle Control**: The final step after a collision free path is planned is to execute the predicted trajectory as closely as possible. This is accomplished via the control inputs such as throttle, brake, and steering [136] that are monitored by closed-loop control algorithms [346]. The proportional-integral-derivative (PID) control algorithm and model predictive control (MPC) algorithm are commonly seen in research and industries [300]. To implement such intelligent control algorithms, the simulator should be capable of building vehicle dynamic models and programming the algorithms in mathematical forms.

5. **3D Virtual Environment**: In order to test various functional elements of a car mentioned in the above requirements, it is equally important to have a realistic

3D virtual environment. The perception system relies on photogenic view of the scene to sense the virtual world. This 3D virtual environment must include both static objects such as buildings, trees, etc. and dynamic objects such as other vehicles, pedestrians, animals, and bicyclists. Furthermore, the dynamic objects must behave realistically to reflect the true behavior of these dynamic entities in an environment.

In order to achieve 3D virtual environment creation, simulators can either rely on game engines or use the high-definition (HD) map of a real environment and render it in a simulation [134]. Similarly, in order to simulate dynamic objects, the vehicle simulators can leverage other domains such as pedestrian models [66] to simulate realistic pedestrians movement in the scene. Furthermore, the 3D virtual environment must support different terrains and weather conditions that are typical in a real environment.

It is important to note that the level of detail in a 3D virtual environment depends on the simulation approach taken. Some companies such as Uber and Waymo do not use highly detailed simulators [134]. Therefore, they do not use simulators to test perception models. However, if the goal is to test perception models in simulation, then the level of detail is very important.

6. **Traffic Infrastructure**: In addition to the requirements for a 3D virtual environment mentioned above, it is also important for a simulation to have the support for various traffic aids such as traffic lights, roadway signage, etc. [282]. This is because these aids help regulate traffic for the safety of all road users. It is projected that the traffic infrastructure will evolve to support connected vehicles in the near future [322]. However, until the connected vehicles become a reality, self-driving cars are expected to comply with the same traffic rules as the human drivers.

7. **Traffic Scenarios Simulation**: The ability to create various traffic scenarios is one of the main points that identify whether a simulator is valuable or not. This allows the researchers to not only recreate/play back a real-world scenario but also allow them to test various "what-if" scenarios that cannot be tested in a real environment because of safety concerns. This criterion considers not only the variety of traffic agents but also the mechanisms that the simulator provides to generate these agents. Different types of dynamic objects consist of humans, bicycles, motorcycles, animals, and vehicles such as buses, trucks, ambulances, and motorcycles. In order to generate scenes close to real-world scenes, it is important that simulator supports significant number of these dynamic agents. In addition, simulator should provide a flexible API that allows users to manage different aspects of simulation, which consists of generating traffic agents and more complex scenarios such as pedestrian behaviors, vehicles crashes, weather conditions, sensor types, stops signs, etc.

8. **2D/3D Ground Truth** In order to provide the training data to the AI models, the simulator should provide object labels and bounding boxes of the objects appearing in the scene. The sensor outputs each video frame where objects are encapsulated in a box.

9. **Non-functional Requirements** The qualitative analysis of open-source simulators includes different aspects that can help AD developers to estimate the
   learning time and the duration required for simulating different scenarios and
   experiments.

   - **Well-Maintained/Stability** In order to use simulator for different experiments
     and testing, the simulator should have comprehensive documentation that
     makes it easy to use. In case that maintenance teams improve the simulator,
     if the backward compatibility is not considered, the documentation should
     provide the precise mapping between the deprecated APIs and newly added
     APIs.
   - **Flexibility/Modular** Open-source simulators should follow division of con-
     cept principle that can help AD developers to leverage and extend different
     scenarios in shorter time. In addition, the simulator can provide a flexible
     API that enables defining customized version of sensors, generating new
     environments, and adding different agents.
   - **Portability** Simulators are expected to run on different types of operating
     systems. Most of the users may not have access to the different types of
     operating systems at the same time; therefore, the simulators portability can
     save the time for the users.
   - **Scalability via a Server Multi-client Architecture** Scalable architecture
     such as client–server architecture enables multiple clients run on different
     nodes to control different agents at the same time. This is helpful specifically
     for simulating the congestion and/or complex scenes.
   - **Open-Source** It is preferred that a simulator be open-source. The open-
     source simulators enable more collaboration, collective progress and allows
     to incorporate learning from peers in the same domain.

6.4 Simulators

This section provides a brief description of simulators that were analyzed and
compared.

6.4.1 MATLAB/Simulink

MATLAB/Simulink published Automated Driving Toolbox™, which provides
various tools that facilitate the design, simulation, and testing of Advanced Driver
Assisted Systems (ADAS) and automated driving systems. It allows users to test
core functionalities such as perception, path planning, and vehicle control. One of its
key features is that HERE HD live map data [204] and OpenDRIVE® road networks
[395] can be imported into MATLAB and these can be used for various design and

testing purposes. Furthermore, the users can build photo-realistic 3D scenarios and model various sensors. It is also equipped with a built-in visualizer that allows to view live sensor detection and tracks [21].

In addition to serving as a simulation and design environment, it also enables users to automate the labeling of objects through the Ground Truth Labeler app [23]. This data can be further used for training purposes or to evaluate sensor performance.

MATLAB provides several examples on how to simulate various ADAS features including Adaptive Cruise Control (ACC), Automatic Emergency Braking (AEB), Automatic Parking Assist, etc. [22]. Last but not least, the toolbox supports Hardware-In-the-Loop (HIL) testing and C/C++ code generation, which enables faster prototyping.

6.4.2 CarSim

CarSim is a vehicle simulator commonly used by industry and academia. The newest version of CarSim supports moving objects and sensors that benefit simulations involving ADAS and Autonomous Vehicles (AVs) [70]. In terms of traffic and target objects, in addition to simulated vehicle, there are up to 200 objects with independent locations and motions. These objects include static objects such as trees and buildings and dynamic objects such as pedestrians, vehicles, animals, and other objects of interest for ADAS scenarios.

The dynamic object is defined based on the location and orientation that is important in vehicle simulation. Additionally, when the sensors are combined with objects, the objects are considered as targets that can be detected.

The moving objects can be linked to 3D objects with their own embedded animations, such as walking pedestrians or pedaling bicyclists. If there are ADAS sensors in the simulation, each object has a shape that influences the detection. The shapes may be rectangular, circular, a straight segment (with limited visibility, used for signs), or polygonal.

In terms of vehicle control, there are several math models available to use and the users can control the motion using built-in options, either with CarSim commands or with external models (e.g., Simulink).

The key feature of CarSim is that it has interfaces to other software such as Matlab and LabVIEW. CarSim offers several examples of simulations and it has a detailed documentation; however, it is not an open-source simulator.

6.4.3 PreScan

PreScan provides a simulation framework to design ADAS and autonomous driving vehicles. It enables manufacturers to test their intelligent systems by providing a variety of virtual traffic conditions and realistic environments. This is benefited by PreScan's automatic traffic generator. Moreover, PreScan enables users to build their customized sensor suites, control logic, and collision warning features.

PreScan also supports Hardware-In-the-Loop (HIL) simulation, which is a common practice for evaluating Electronic Control Unit (ECU). PreScan is good at physics-based calculations of the sensor inputs. The sensor signals are input to the ECU to evaluate various algorithms. Furthermore, the signals can also be output to either the loop or camera HIL for driver. It also supports real-time data and GPS vehicle data recording, which can then be replayed later on. This is very helpful for situations which are otherwise not easy to simulate with synthetic data.

Additionally, PreScan offers a unique feature called Vehicle Hardware-In-the-Loop (VeHIL) laboratory. It allows users to create a combined real and virtual system where the test/ego vehicle is placed on a roller bench and other vehicles are represented by wheeled robots with a car-like appearance. The test vehicle is equipped with realistic sensors. By using this combination of ego vehicle and mobile robots, VeHIL is capable of providing detailed simulations for ADAS.

6.4.4 CARLA

CARLA [119] is an open-source simulator that democratizes autonomous driving research area. The simulator is open-source and is developed based on the Unreal Engine [530]. It serves as a modular and flexible tool equipped with a powerful API to support training and validation of ADAS systems. Therefore, CARLA tries to meet the requirements of various use cases of ADAS, for instance, training the perception algorithms or learning driving policies. CARLA is developed from scratch based on the Unreal Engine to execute the simulation and it leverages the OpenDRIVE standard to define roads and urban settings. The CARLA API is customizable by users and provides control over the simulation. It is based on Python and C++ and is constantly growing concurrently with the project that is an ecosystem of projects, built around the main platform by the community.

The CARLA consists of a scalable client–server architecture. The simulation-related tasks are deployed at the server including the updates on the world-state and its actors, sensor rendering, computation of physics, etc. In order to generate realistic results, the server should run with the dedicated GPU. The client side consists of some client modules that control the logic of agents appearing in the scene including the pedestrians, vehicles, bicycles, and motorcycles. Also, the client modules are responsible for the world conditions setups. The setup of all client modules is achieved by using the CARLA API. The vehicles, buildings, and urban layouts are couple of open digital assets that CARLA provides. In addition, the environmental conditions such as different weather conditions and flexible specification of sensor suits are supported. In order to accelerate queries (such as the closest waypoint in a road), CARLA makes use of RTrees.

In recent versions, CARLA has more accurate vehicle volumes and more realistic core physics (such as wheel's friction, suspension, and center of mass). This is helpful when a vehicle turns or a collision occurs. In addition, the process of adding traffic lights and stop signs to the scene have been changed from manual to automatic by leveraging the information provided by the OpenDRIVE file.

CARLA proposes a safety assurance module based on the RSS library. The responsibility of this module is to put holds on the vehicle controls based on the sensor information. In other words, the RSS defines various situations based on sensor data and then determines a proper response according to safety checks. A situation describes the state of the ego vehicle with an element of the environment. Leveraging the OpenDrive signals enables the RSS module to take different road segments into consideration that helps to check the priority and safety at junctions.

6.4.5 Gazebo

Gazebo is an open-source, scalable, flexible, and multi-robot 3D simulator [267]. It is supported on multiple operating systems, including Linux and Windows. It supports the recreation of both indoor and outdoor 3D environments.

Gazebo relies on three main libraries, which include physics, rendering, and a communication library. Firstly, the physics library allows the simulated objects to behave as realistically as possible to their real counterparts by letting the user define their physical properties such as mass, friction coefficient, velocity, inertia, etc. Gazebo uses Open Dynamic Engine (ODE) as its default physics engine, but it also supports others such as Bullet, Simbody and Dynamic Open-Source Physics Engine (DART). Secondly, for visualization, it uses a rendering library called Object-Oriented Graphics Rendering Engine (OGRE), which makes it possible to visualize dynamic 3D objects and scenes. Thirdly, the communication library enables communication among various elements of Gazebo. Besides these three core libraries, Gazebo offers plugin support that allows the users to communicate with these libraries directly.

There are two core elements that define any 3D scene. In Gazebo terminology, these are called a world and a model. A world is used to represent a 3D scene, which could be an indoor or an outdoor environment. It is a user-defined file in a Simulation Description File (SDF) format [474], with a dot world extension. The world file consists of one or many models. Furthermore, a model is any 3D object. It could be a static object such as a table, house, sensor, or a robot or a dynamic object. The users are free to create objects from scratch by defining their visual, inertial, and collision properties in an SDF format. Optionally, they can define plugins to control various aspects of simulation, such as, a world plugin controls the world properties, and model plugin controls the model properties, and so on. It is important to note that Gazebo has a wide community support that makes it possible to share and use models already created by somebody else. Additionally, it has well-maintained documentation and numerous tutorials.

Finally, Gazebo is a standalone simulator. However, it is typically used in conjunction with ROS [503, 579]. Gazebo supports modeling of almost all kinds of robots. The work [69] presents a complex scenario that shows the advanced modeling capabilities of Gazebo which models Prius Hybrid model of a car driving in the simulated M-city.

6.4.6   LGSVL

LG Electronics America R&D Center (LGSVL) [446] is a multi-robot AV simulator. It proposes an out-of-the-box solution for the AV algorithms to test the autonomous vehicle algorithms. It is integrated to some of the platforms that make it easy to test and validate the entire system. The simulator is open-source and is developed based on the Unity's game engine [529]. LGSVL provides different bridges for message passing between the AD stack and the simulator backbone.

The simulator has different components. The user AD stack provides the development, test, and verification platform to the AV developers. The simulator supports ROS1, ROS2, and Cyber RT messages. This helps to connect the simulator to the Autoware [20] and Baidu Apollo [29], which are the most popular AD stacks. In addition, multiple AD simulators can communicate simultaneously with the simulator via ROS and ROS2 bridges for the Autoware and customized bridge for Baidu Apollo. LGSVL Simulator leverages Unity's game engine that helps to generate photo-realistic virtual environments based on High-Definition Render Pipeline (HDRP) technology. The simulation engine provides different functions for simulating the environment simulation (traffic simulation and physical environment simulation), sensor simulation, and vehicle dynamics. The simulator provides a Python API to control different environment entities. In addition, sensor and vehicle models propose a customizable set of sensors via setting up a JSON file that enables the specification of the intrinsic and extrinsic parameters. The simulator currently supports camera, LiDAR, IMU, GPS, and radar. Additionally, developers can define customized sensors. The simulator provides various options, for instance, the segmentation and semantic segmentation. Also, LGSVL provides Functional Mockup Interface (FMI) in order to integrate vehicle dynamics model platform to the external third-party dynamics models. Finally, the weather condition, the day time, traffic agents, and dynamic actors are specified based on 3D environment. One of the important features of LGSVL is exporting HD maps from 3D environments.

6.5   Comparison

This section provides a comparison of different simulators described under Sect. 6.4, starting with MATLAB, CarSim, and PreScan. Then we compare Gazebo and

CARLA, followed by comparison of CARLA and LGSVL. Finally, we conclude with our analysis and key observations.

MATLAB/Simulink is designed for the simple scenarios. It is good at computation and has efficient plot functions. The capability of co-simulation with other software like CarSim makes it easier to build various vehicle models. It is common to see users using the vehicle models from CarSim and build their upper control algorithms in MATLAB/Simulink to do a co-simulation project. However, MATLAB/Simulink has limited ability to realistically visualize the traffic scenarios, obstacles, and pedestrian models. PreScan has strong capability to simulate the environment of the real world such as the weather conditions that MATLAB/Simulink and CarSim cannot do, and it also has interfaces with MATLAB/Simulink that makes modeling more efficient.

Furthermore, Gazebo is known for its high flexibility and its seamless integration with ROS. While the high flexibility is advantageous because it gives full control over simulation, it comes at a cost of time and effort. As opposed to CARLA and LGSVL simulators, the creation of a simulation world in Gazebo is a manual process where the user must create 3D models and carefully define their physics and their position in the simulation world within the XML file. Gazebo does include various sensor models and it allows users to create new sensor models via the plugins. Next, we compare CARLA and LGSVL simulators.

Both CARLA and LGSVL provide high-quality simulation environments that require GPU computing unit in order to run with reasonable performance and frame rate. The user can invoke different facilities in CARLA and LGSVL via using a flexible API. Although, the facilities are different between two simulators. For instance, CARLA provides build-in recorder while LGSVL does not provide. Therefore, in order to record videos in LGSVL, the user can leverage video recording feature in Nvidia drivers. CARLA and LGSVL provide a variety of sensors, some of these sensors are common between them such as Depth camera, LiDAR, and IMU. In addition, each simulator provides different sensors with description provided in their official website. Both simulators, CARLA and LGSVL, enhance users to create the custom sensors. The new map generation has different processes in CARLA and LGSVL. The backbone of CARLA simulator is Unreal Engine that generates new maps by automatically adding stop signs based on the OpenDRIVE technology. On the other hand, the backbone of LGSVL simulator is Unity's game engine and the user can generate new map by manually importing different components into the Unity's game engine. Additionally, the software architecture in CARLA and LGSVL is quite different. LGSVL mostly connects to AD stacks (Autoware, Apollo, etc.) based on different bridges, and most of the simulators' facilities publish or subscribe the data on specified topics in order to enable AD stacks to consume data. On the other hand, most of facilities in CARLA are built in, although it enables users to connect to ROS1, ROS2, and Autoware via using the bridges.

While all of the six simulators described in the chapter offer their own advantages and disadvantages, we make the following key observations:

- *Observation 1:* LGSVL and CARLA are most suited for end-to-end testing of unique functionalities that self-driving cars offer such as perception, mapping, localization, and vehicle control because of many built-in automated features they support.
- *Observation 2:* Gazebo is a popular robotic simulator, but the time and effort needed to create dynamic scenes do not make it the first choice for testing end-to-end systems for self-driving cars.
- *Observation 3:* MATLAB/Simulink is one of the best choices for testing upper-level algorithms because of the clearly presented logic blocks in Simulink. Additionally, it has a fast plot function that makes it easier to do the analysis of results.
- *Observation 4:* CarSim specializes in vehicle dynamic simulations because of its complete vehicle library and variety of vehicle parameters available to tune. However, it has limited ability to build customized upper-level algorithms in an efficient way.
- *Observation 5:* PreScan has a strong capability of building realistic environments and simulating different weather conditions.

In Table 6.2, we provide a comparison summary where all six simulators described in this chapter are further compared.

6.6 Challenges

The automotive simulators have come a long way. Although simulation has now become a cornerstone in the development of self-driving cars, common standards to evaluate simulation results are lacking. For example, the Annual Mileage Report submitted to the California Department of Motor Vehicle by the key players such as Waymo, Cruise, and Tesla does not include the sophistication and diversity of the miles collected through simulation [381]. It would be more beneficial to have simulation standards that could help make a more informative comparison between various research efforts.

Furthermore, we are not aware of any simulators that are currently capable of testing the concept of connected vehicles, where vehicles communicate with each other and with the infrastructure. However, there are testbeds available such as the ones mentioned in the report [120] from the US Department of Transportation.

In addition, current simulators, for instance, CARLA and LGSVL, are on-going projects and add the most recent technologies. Therefore, the user may encounter with undocumented errors or bugs. Therefore, the community support is quite important which can improve the quality of open-source simulators and ADAS tests.

Table 6.2 Comparison of various simulators

Requirements	MATLAB MATLAB	CarSim	PreScan	CARLA	Gazebo	LGSVL
Perception (sensor models)	Y	Y	Y	Y(1)	Y(2)	Y(3)
Perception (different weather conditions)	N	N	Y	Y	N	Y
Camera calibration	Y	N	Y	Y	N	N
Path planning	Y	Y	Y	Y	Y	Y
Vehicle control (vehicle dynamics)	Y	Y	Y	Y	Y	Y(3)
3D virtual environment	U	Y	Y	Y, Outdoor (urban)	Y, indoor and outdoor	Y, outdoor (urban)
Traffic infrastructure	Y, with traffic lights	Y	Y	Y, Traffic lights, intersections, stop signs, lanes	Y	Y
Traffic scenario simulation (dynamic objects)	Y	Y	Y	Y	N(2)	Y
2D/3D ground truth	Y	N	N	Y	U	Y
Interfaces to other software	Y, with CarSim, PreScan, ROS	Y, with Simulink	Y, with Simulink	Y, with ROS, Autoware	Y, with ROS	Y, with Autoware, Apollo, ROS
Scalability	U	U	U	Y	Y	Y
Open-source	N	N	N	Y	Y	Y
Well maintained	Y	Y	Y	Y	Y	Y
Portability	Y	Y	Y	Y	Y	Y
Flexible API	Y	Y	U	Y (2)	Y	Y

6.7 Summary

In this chapter, we compare MATLAB/Simulink, CarSim, PreScan, Gazebo, CARLA, and LGSVL simulators for testing self-driving cars. The focus is on how well they are at simulating and testing perception, mapping and localization, path planning, and vehicle control for the self-driving cars. Our analysis yields five key observations that are discussed in Sect. 6.5. We also identify key requirements that state-of-the-art simulators must have to yield reliable results. Finally, several challenges still remain with the simulation strategies such as the lack of common standards as mentioned in Sect. 6.6. In conclusion, simulation will continue to help design self-driving vehicles in a safe, cost effective, and timely fashion, provided the simulations represent the reality.

Chapter 7
Hardware Platforms

7.1 HydraOne

7.1.1 Introduction

There are numerous academic studies and industrial works of edge computing that have emerged in the past few years, crossing various domains [461, 483]. Many researchers and developers have focused more attention on critical edge applications [313, 542], framework and middleware for edge computing [593], security and consistency on edge [364], and IoT edge–cloud interactions [505], etc. However, few studies have discussed how to develop a research platform for edge computing in a specific application scenario, which is more important for sharing and spreading research achievements in the edge computing field.

In the cloud computing domain, the main function of the research platform is data processing. Researchers focus more on computing performance when building the research platform for cloud computing. The development of virtualization technology [388] and distributed computing [100] allowed researchers to build their private cloud computing platform via some open-source framework [477]. In the Internet of Things (IoT) domain, the research platform is used to collect and transmit sensor data, so researchers are more concerned about the peripheral interface resources and wireless communication capability of the research platform. Many classic IoT platforms for research and education have been released, such as, Arduino [32] and Telos [417]. The edge computing domain has the characteristics of both cloud computing and the IoT domain. The research platform for edge computing should have enough computational capabilities, and it could collect the data from the data producers in specific computing scenarios and communicate with other entities in the network.

Connected and autonomous vehicles (CAVs) are already changing our vision about future vehicles and transportation. A recent report shows that each connected and autonomous vehicle will generate about 4000 GB of data per day [277]. The

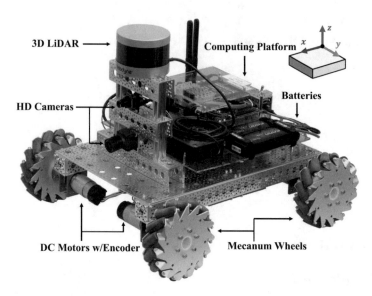

Fig. 7.1 Overview of HydraOne platform

majority of vehicular data would be processed in the vehicle due to network bandwidth restrictions under which CAVs become a typical edge computing system. A large number of research opportunities still remain in the field of edge computing in CAVs. However, it is challenging for researchers to deploy applications and systems designed for CAVs in real-world environments due to the lack of a research platform for CAVs. To address this challenge, we propose HydraOne, an indoor experimental research and education platform for edge computing in the CAVs scenario. As shown in Fig. 7.1, HydraOne is a full-stack research and education platform from hardware to software, including mechanical components and vision sensors, and a computing and communication system. All the resources on HydraOne are managed by the Robot Operating System (ROS) [424]. HydraOne has three key characteristics: design modularization, resource extensibility and openness, as well as function isolation, which allows users to conduct various research and education experiments on CAVs with HydraOne.

   While HydraOne is an indoor robot-based platform, it has sufficient resources and components to conduct CAVs experiments. The computing platform on HydraOne collects and processes the sensor data in real-time; then, it outputs the control message to the chassis to control the moving speed and direction of the platform, which enables the autonomous driving capability of HydraOne. The users can develop the algorithms of sensor fusion, perception, and decision-making on the platform. HydraOne is equipped with a WiFi module to communicate with the cloud and the edge server, which allows users to conduct the experiments of Vehicle-to-Everything (V2X) on HydraOne. In addition to autonomous driving and communication, the research problems supported by HydraOne include but are not limited to the operating system designed for CAVs, safe and trusted execution environments on CAVs, and privacy preservation model and tools for vehicular data.

7.1.2 Related Work

In this section, we summarize the related work from the perspective of two research platforms: for autonomous devices and edge computing.

Research Platform for Autonomous Devices Wei et al. presented the CMU autonomous driving research platform, which is based on a Cadillac SRX [557]. This work focuses on vehicle engineering problems, including the actuation, power, and sensor systems on the vehicle. Tomic et al. presented an autonomous UAV research platform for urban search and rescue [523]. They introduced the hardware infrastructure of their platform and provided a set of algorithm libraries to help developers complete the urban search and rescue task. Pheeno [560] and r-one [347] both are research and education platform for multi-robot manipulation. These studies designed a low-cost robot platform with a small number of sensors and a low-power communication module, which can help researchers to deploy the experiment of versatile swarm robotic.

Research Platform for Edge Computing ParaDrop [315] is an edge computing platform designed for multi-tenant on wireless gateways. It uses Linux Container (LXC) to manage the resource on the wireless gateway, which allows researchers to implement their edge computing applications on the gateway. Φ-stack [572] is a full-stack research platform for the smart web of things. Φ-stack contains a novel smart IoT processor (ΦPU) and a RESTful-based software framework (ΦOS and ΦDK), which natively supports the web and intelligence. Researchers can use Φ-stack to deploy the intelligence workloads via RESTful API on low-power smart IoT devices.

It should be noted that there some similar platforms already exist in the community, such as MIT RACECAR [250], DJI RoboMaster Robots [114], and Audi Autonomous Model Cars [2], but they only can be acquired by participating in specific competitions. HydraOne provides an open platform for researchers and students to be able to build their own CAVs experimental platform according to this chapter.

7.1.3 Design and Implementation

In this section, we introduce the design and implementation details of HydraOne and present three key characteristics of the platform.

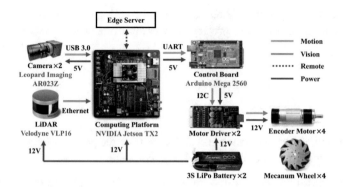

Fig. 7.2 HydraOne hardware design

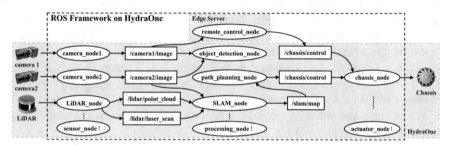

Fig. 7.3 HydraOne software framework

7.1.3.1 Hardware Design

Design Overview As shown in Fig. 7.2, HydraOne is equipped with an NVIDIA®
Jetson™ TX2 embedded module [391] as the core computing platform. Jetson TX2
collects the data from multiple sensors and feeds the data to several computing
tasks in real-time. An Arduino board receives the control message output from the
computing tasks via serial port (UART) and then sends the control signals to the
motor drivers to control the movement of HydraOne. Jetson TX2 is equipped with a
WiFi module, which allows HydraOne to communicate with an edge server or other
entities in the network. The whole HydraOne platform is powered by two 3S LiPo
batteries.

Sensors HydraOne is equipped with two HD cameras and one 3D LiDAR, which
form the basic vision system of HydraOne. The sensors' models are listed as
below:

- 2×Leopard Imaging AR023Z 1080p HD color camera, 1920 × 1080@30, rolling
  shutter;
- 1×Velodyne VLP-16 rotating 3D laser scanner, 16 channels, collecting ∼ 3
  million points/second, field of view: 360° horizontal, 30° vertical, range: 100 m.

Computing Platform The computing platform on HydraOne processes complicated computing tasks, such as computer vision and machine learning algorithms, to support the various applications of CAVs. The traditional microprocessor cannot meet the computing power requirement of the CAVs scenario, so we should deploy a single board computer on the HydraOne platform. NVIDIA® Jetson™ TX2 is a power-efficient embedded AI computing platform which has dual-core Denver and quad-core ARM® Cortex-A57 CPU equipped with 8GB DDR4 memory and 32GB eMMC. The GPU on Jetson TX2 is powered by NVIDIA Pascal™ architecture with 256 CUDA cores. The CPU-GPU architecture on Jetson TX2 can accelerate the deep learning workloads, which have become an integral part of the CAVs scenario [305, 549]. Several studies have deployed the Jetson TX2 on autonomous devices [250, 521]. Therefore, Jetson TX2 is currently the most suitable choice for a computing platform on HydraOne.

Chassis The HydraOne platform has a four-wheel drive chassis which is equipped with four DC motors with encoders and two encoder motor drivers. The proportional-integral-derivative (PID) control algorithm is integrated into the motor drivers to achieve precision control for each motor speed. The chassis has four Mecanum wheels (a kind of Omni wheel) so that the HydraOne can achieve omnidirectional movement. The control message format output from computing platform is *geometry_msgs/Twist* (this will be introduced in the next subsection); the Arduino board is in charge of converting the message to motor speed value and sending the speed value to the motor drivers via I2C bus.

Power System The electronic components and chassis have an independent power supply. Each is powered by one 3S LiPo battery. We present the running power breakdown of HydraOne in Fig. 7.4. The three applications will be introduced in Sect. 7.1.4. The results show that the whole platform consumes $41.2w$ on average when running workloads, and sensors, computing platform, and chassis consume $11.2w$ (27.2%), $8.1w$ (19.7%), and $21.9w$ (52.2%), respectively.

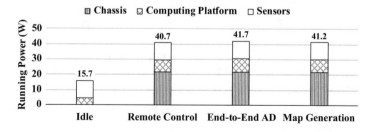

Fig. 7.4 Running power breakdown on HydraOne

7.1.3.2 Software Framework

Framework Overview The operating system on Jetson Tx2 is Ubuntu 16.04, so the users can easily install the open-source vision and machine learning libraries, like OpenCV, TensorFlow, PCL, etc. To manage the hardware and software resources on HydraOne and provide a clear and easy-to-use development model to researchers, we deploy the Robot Operating System (ROS) [424] on HydraOne. We choose ROS Kinetic Kame distribution, which is the most compatible version to Ubuntu 16.04 to date. The ROS framework on HydraOne is illustrated in Fig. 7.3. All resources and computing tasks on HydraOne can be abstracted as ROS nodes, and they use the publisher–subscriber pattern to share data and results to implement one or several CAVs applications collectively. The communication between HydraOne and the edge server is also implemented by ROS. HydraOne runs the ROS master node, and the edge server should configure the IP address and port of the master node in its environments.

Node Management The ROS nodes on HydraOne are divided into three categories according to their function: sensor node, processing node, and actuator node. The sensor nodes are data producers that collect the data from hardware sensors and publish them in real-time. It must be noted that one sensor node could publish multiple types of data, and all data producers can be considered as sensor nodes, such as the motor speed monitor node, the battery status monitor node, etc. The processing nodes are the instantiation of edge computing on HydraOne, so all CAVs computing workloads should be implemented in the processing nodes. Some processing nodes will publish the middle results to other nodes, and some will publish the control message to actuator nodes. The actuator nodes are in charge of controlling the hardware actuator (e.g., chassis motors) to make correct and prompt responses according to the control message. The actuator can receive the message from multiple processing nodes, so the users should manage the control priority when more than one processing node is publishing the control message to the same actuator node.

Message Flow An ROS message is essentially an implementation of inter-process communication (IPC). ROS nodes can pass the sensor data, processing results, and control message via the ROS message to others. The execution process of the CAVs application can be regarded as message passing and processing among the ROS nodes, which can be abstracted as message flow. Some properties of message flow in our framework are summarized as follows:

- An application of CAVs deployed on HydraOne can be abstracted as one or several message flows.
- Message flow consists of messages and nodes, messages connect nodes, and nodes pass or transform messages.
- Ordinarily, the message flow starts from sensor messages and ends with control messages to actuator nodes.

Development Model The development model of HydraOne is based on the node management method and message flow abstraction we mentioned above. We provide sensor nodes of the vision sensors on HydraOne and the actuator node of the HydraOne chassis. The sensor data ROS format is *sensor_msgs/Image* (cameras), *sensor_msgs/PointCloud2* (LiDAR-3D), and *sensor_msgs/LaserScan* (LiDAR-2D). The format of the control message to the chassis node is *geometry_msgs/Twist*, which contains two 3-tuple vectors indicating the linear and angular speed in x-, y-, and z-axis, respectively. The users can focus on developing the CAVs applications and algorithms on processing nodes, just subscribe the data from the sensor nodes to feed into their CAVs tasks, and output the control message to control the movement of HydraOne. The development model is clear and concise, which allows researchers to test and evaluate their CAVs applications and algorithms in real-world environments easily and quickly.

7.1.3.3 Experimental Enablers

The HydraOne platform has three key characteristics: design modularization, resource extensibility and openness, and function isolation. Understanding the three key characteristics of HydraOne will help users take full advantage of the platform to conduct research and education experiments of edge computing on CAVs.

Design Modularization The idea of modular design is inspired by LEGO® robot [186]; all the hardware modules are connected via standard interfaces, so the users can easily test, replace, and upgrade each module. The ROS node and message are the implementation of the modular design of the software framework on HydraOne. Every node has a limited function and is connected via standard interfaces (messages). The design modularization will help users learn every module on HydraOne on both the hardware and software levels and fully understand the development model of the platform.

Resource Extensibility and Openness Based on the modular design, the HydraOne platform is resource extensible. The structural components of HydraOne allow users to easily mount new hardware resources on the platform, and the users need to provide the driver node of each resource to publish or subscribe to data resources. All the platform resources are open. In addition to the development model we provide, the users can access, analyze, and customize any resources on HydraOne, even replace the whole software framework. The resource openness allows users to explore some system and architecture research problems on HydraOne, such as, designing an operating system for CAVs.

Function Isolation Function isolation is reflected in two aspects: within the framework and as the framework is shared with other libraries. The sensor nodes and actuator nodes are responsible for managing hardware resources, and each node only manages one hardware module. The processing nodes do not access the hardware directly to prevent exclusive access to hardware resources. The processing nodes will call other function libraries to complete the computing tasks. Some libraries provide their own process manage function, like *session.run()* in TensorFlow. We recommend that the users use ROS to manage each node (process) to isolate the ROS function with other libraries and only use the standard interface to call them. The function isolation will make it easier for users to program and debug on the HydraOne platform.

7.1.4   Case Studies

In this section, we use three case studies deployed on HydraOne to show the capabilities of our research platform.

7.1.4.1   Remote Control

Currently, autonomous driving systems are based on computer vision and machine learning algorithms, which cause failure in some unrecognized situations [243]. However, as it is extremely hard for training datasets to cover all circumstances of the real environment, autonomous driving system failure is unavoidable. Therefore, the remote control is an essential application to guarantee the safety of autonomous driving vehicles.

The message flow of remote control is shown in Fig. 7.5. The remote control node is a processing node launched on the edge server. It subscribes the *sensor_msgs/Image* message, which is published by camera node, and displays the images to let the operator know the HydraOne status. The operator sends the *geometry_msgs/Twist* message via keyboard or other controllers according to the HydraOne status. The chassis node subscribes the control message to adjust the running speed and direction of HydraOne.

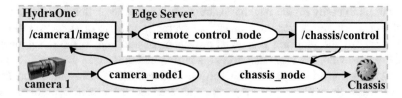

Fig. 7.5 Message flow of remote control

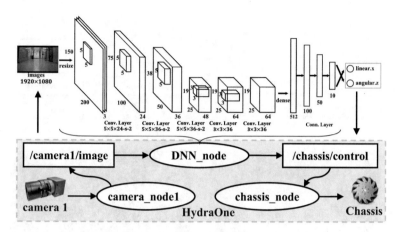

Fig. 7.6 Message flow of end-to-end autonomous driving

7.1.4.2 Autonomous Driving

As a CAVs research platform, supporting autonomous driving is one of the core functions. Inspired by some end-to-end deep neural network (DNN) algorithms (e.g., SSD [320] and YOLO [434]) and DAVE-2 [52], an end-to-end system for self-driving cars, we deploy a DNN-based end-to-end autonomous driving application on HydraOne.

The message flow of end-to-end autonomous driving is shown in Fig. 7.6. The DNN node is a processing node launched on HydraOne. It subscribes the *sensor_msgs/Image* message and takes the resized images as the input of the DNN model. The output of the model is the linear speed on *x*-axis and angular speed on *z*-axis, which are fed into *geometry_msgs/Twist* message to control the chassis. The DNN model consists of five convolutional layers and four fully connected layers, and we chose ReLU as the activation function for all layers. Other details of the model, like the convolution kernel and stride size, are illustrated in Fig. 7.6.

We use a joystick to control HydraOne when collecting the training data. The message data is recorded by rosbag file. The *geometry_msgs/Twist* message is the label of images, and we use the timestamp to match them. We train the model on a GPU server and download the model to the edge (HydraOne) to process the data. The training and inference process is implemented on the TensorFlow framework.

7.1.4.3 Map Generation

The indoor map will help HydraOne have a better understanding of the surrounding environment. Furthermore, a complete indoor map is the base data of some upper-level CAVs applications, such as path planning. We use the LiDAR data to implement an indoor mapping case study on HydraOne.

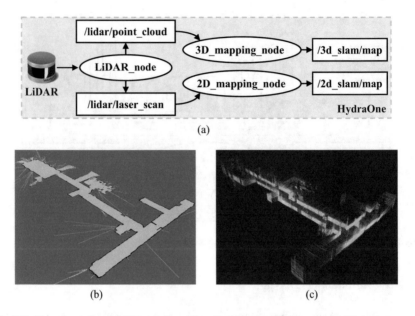

(a)

(b)   (c)

Fig. 7.7 Map generation. (**a**) Message flow of map generation. (**b**) 2D map. (**c**) 3D map

The message flow of indoor mapping is shown in Fig. 7.7a. The LiDAR node published 2D laser scan data *sensor_msgs/LaserScan* and 3D point cloud data textitsensor_msgs/PointCloud2. The mapping nodes subscribe to the LiDAR data and publish 2D and 3D map messages, respectively. The demo of the map data is presented in Fig. 7.7b, c. The indoor mapping function usually runs in conjunction with the remote control or free space detection nodes to generate the map in new environments.

7.1.5 Summary

In this chapter, we present the design, implementation, characteristics, and case studies of HydraOne, the indoor experimental research and education platform for edge computing in the CAVs scenario. HydraOne is modularly designed; the resources (hardware and software) on HydraOne are extensible and all open to users. In addition, all the function modules on HydraOne are isolated. These three characteristics will allow users to take full advantage of the HydraOne to conduct research and education experiments of edge computing on CAVs. The research problems for CAVs supported by HydraOne are not only limited to algorithms and applications for autonomous driving and V2X but also include full-stack system-related problems, such as hardware platform and architecture, operating system and software middleware, security and privacy, etc. We hope this platform will be valuable to researchers and students who are working on edge computing in connected and autonomous vehicles.

7.2 Hydra

An autonomous vehicle involves multiple subjects, including computing systems, machine learning, communication, robotics, mechanical engineering, and systems engineering, to integrate different technologies and innovations. Figure 7.8 shows a typical example of autonomous driving vehicles called *Hydra*, which is developed by The CAR lab at Wayne State University [516]. An NVIDIA Drive PX2 is used as the vehicle computation unit (VCU). Multiple sensors, including six cameras, six radars, one LiDAR, one GNSS antenna, and one DSRC antenna are installed for sensing and connected with VCU. The CAN bus is used to transmit messages between different ECUs controlling steering, throttle, shifting, brake, etc. Between the NVIDIA Drive PX2 and the vehicle's CAN bus, a drive-by-wire system is deployed as an actuator of the vehicle control commands from the computing system. Additionally, a power distribution system is used to provide extra power for the computing system. It is worth noting that the computing system's power distribution is non-negligible in modern AVs [309].

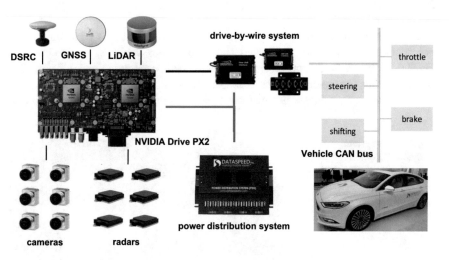

Fig. 7.8 A typical example of a computing system for autonomous driving

7.3   Equinox: A Road-Side Edge Computing Experimental Platform for CAVs

7.3.1   Introduction

Owing to the high safety and efficiency, connected and autonomous vehicle (CAV) has become the fundamental technology for the next generation of transportation [548, 600]. According to the estimation from Intel, with complicated sensors including cameras, radars, LiDAR, etc. installed on the vehicle, the total amount of data generated on one vehicle can attain 4000 TB per day. The challenge of processing this huge amount of data efficiently pushes the services originally provided on the cloud to the edge of the network [483]. Therefore, both the vehicle and the Road-Side Unit (RSU) have been implemented with powerful computing platforms; how to leverage the computing systems together for connected and autonomous driving applications is a remaining issue.

Recently, lots of works have been working on how to leverage the computing resources on the RSU for the applications running on top of the vehicle. Numerous research works and projects are focused on how to schedule the offloading of tasks onto the RSU [144, 546]. However, these works are not implemented and evaluated on real RSU prototype. Another portion of research works aim at designing an initial prototype to realize RSU platform [5]. However, they are all based on single communication method. In summary, there is no standard design as well as the prototype of the RSU.

In this chapter, we proposed Equinox, which is a road-side edge computing experimental platform for connected and autonomous vehicles. As is shown in Fig. 7.9, Equinox is composed of three main layers: communication later, data layer, and the computation layer. The communication layer enables Equinox to communicate with vehicles in three ways: WiFi, Dedicated Short Range Commu-

Fig. 7.9 The system overview of Equinox

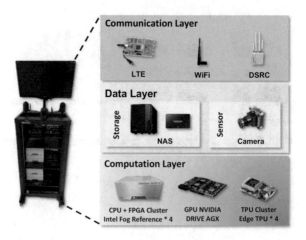

nication (DSRC), and LTE. Since huge amount of data could be uploaded and stored on the RSU, the data layer is proposed to include the data generation and data storage into the system of Equinox. Hence, a network-attached storage (NAS) and several solid-state drives (SSDs) are designed to be included in Equinox. In addition, Equinox is supposed to perform sensing and perception based on its own sensors. Through this design, Equinox not only provides computation resources for offloading tasks running on the vehicle but also broadcasts traffic information through its own sensing and perception process. Last but not least, Equinox is designed with a heterogeneous computation layer, in which a CPU cluster with Field Programmable Gate Array (FPGA) accelerator, GPU, as well as edge Tensor Processing Unit (TPU). This heterogeneous computing platform is used to provide efficient processing of CAV applications especially machine learning-based algorithms.

7.3.2 System Design

The goal of Equinox is to provide not only resources like computation and storage but also the traffic information that is essentially extracted through the sensing and perception of Equinox. From Fig. 7.9, we understand the function and objective of each layer. In this section, firstly we discuss the overall design of Equinox, followed by the detailed design of each layer.

7.3.2.1 Communication Layer

The communication layer is supposed to provide reliable and sufficient commu-nication between the vehicle and RSU, which can be represented as vehicle to infrastructure (V2I) in short. To the best of our knowledge, the majority of the-state-of-the-art works focused on leverage the V2I communications for the sharing of important warning information for the safety purpose. The biggest challenge for these research works is how to provide reliable bandwidth even in mobile environment. For example, DSRC is able to provide reliable V2X communications even when the vehicle is driving 120 km/h, but the communication coverage and bandwidth are limited. In contract, WiFi-based communication provides high bandwidth but is also limited by the coverage, while LTE-based communication does not has the coverage and bandwidth limitation, but neither of them can provide stable communication in mobile scenario.

Therefore, the communication layer of Equinox combines these three types of communication mechanism together. As a complimentary to one another, Equinox achieves reliable and sufficient V2I communication. The choice of which commu-nication mechanism to use is based on the whether the application is data-driven or latency-driven, as well as the state of the vehicle and Equinox.

In terms of implementation, all the computing boards within Equinox are equipped with WiFi antenna. An NI USRP-based board is deployed on Equinox to provide LTE based communications, while OpenAirInterface-based software makes it can be programmable for higher level applications. For DSRC, a Road-Side Entity (RSE) with an On-Board Entity (OBD) is deployed on Equinox and vehicle, respectively. A 24-port Ethernet switch with a router is deployed in Equinox to make all the communication, storage, and computing devices connected and accessible to one another. RSE is also connected to the switch.

7.3.2.2 Data Layer

Driven by the goal of supporting not only data and computation offloading but also traffic information perception and broadcasting, there can be huge amount of variety of data generated and processed on Equinox. How to store the data efficiently on Equinox is the first challenge. In our proposed design, a network-attached storage (NAS), which has five 10T of hard disk and five 1T of SSD, will be deployed on Equinox. NAS is also connected to the Ethernet switch to make it accessible to other components. In terms of NAS, another challenge is how to manage the data, since some data should be open to developers in a security and privacy preserving way, while some data can be out of date and should be compressed or even permanently deleted.

Besides NAS, another important part of the data layer is the sensors. One of the biggest reasons we also consider sensors on the RSU is to make it capable of sensing and perception of the environment because the RSU is usually installed fixed and it gets a better view on the traffic compared with the vehicle. Making the RSU become intelligent is the key to leverage the traffic infrastructures to improve the vehicle's safety. On Equinox, currently we installed three surveillance cameras which get a bird's-eye view of the indoor autonomous driving robots.

7.3.2.3 Computation Layer

As an essential portion of Equinox, the computation layer is responsible for the sensing and perception of Equinox's environment, the execution of tasks offloaded from the vehicle, and the management of the communication as well as the storage system. Considering the percentage of machine learning especially deep learning-based applications, a heterogeneous computation platform is proposed in Equinox.

The proposed heterogeneous computation platform includes four Inter Fog Reference-based CPU-FPGA cluster, four Edge TPU-based cluster, and an NVIDIA GPU. In particular, each Intel Fog Reference has Xeon E3 CPU and Cyclone V FPGA board. The memory size of Intel Fog Reference is 32 GB. Edge TPU is designed to accelerate the model inference on the edge. Each edge TPU board has an Integrated GC7000 GPU, 1 GB RAM and 8 GB of flash memory. The NVIDIA

GPU is NVIDIA DRIVE AGX Xavier, which is the newest version for autonomous driving vehicle and we use it to accelerate both model training and inference.

All the computing devices are connected to the Ethernet switch. For the software part, all the devices are Linux-based system, and we use Spark to manage the application running in the whole system and ROS to manage data sharing process between the vehicle and Equinox.

Chapter 8
Smart Infrastructure for Autonomous Driving

8.1 Innovations on the V2X Infrastructure

One effective method to alleviate the huge computing demand on autonomous driving edge computing systems is Vehicle-to-Everything (V2X) technologies. V2X is defined as a vehicle communication system, which consists of many types of communications: vehicle-to-vehicle (V2V), vehicle-to-network (V2N), vehicle-to-pedestrian (V2P), vehicle-to-infrastructure (V2I), vehicle-to-device (V2D), and vehicle-to-grid (V2G). Currently, most research focuses on V2V and V2I. While conventional autonomous driving systems require costly sensors and edge computing equipment within the vehicle, V2X takes a different and potentially complimentary approach by investing on road infrastructure, thus alleviating the computing and sensing costs in vehicles.

As the rapid deployment of edge computing facilities in the infrastructure, more and more autonomous driving applications have started leveraging V2X communications to make the in-vehicle edge computing system more efficient. One popular direction is cooperative autonomous driving. The cooperation of autonomous driving edge computing system with V2X technology makes it possible to build a safe and efficient autonomous driving system [496]. However, the application and deployment of cooperative autonomous driving systems are still open research problems. In this section, we discuss the evolution of V2X technology and present some case studies of V2X for autonomous driving: convoy driving, cooperative lane change, cooperative intersection management, and cooperative sensing.

W. Shi, L. Liu, *Computing Systems for Autonomous Driving*,
https://doi.org/10.1007/978-3-030-81564-6_8

8.1.1   The Evolution of V2X Technology

In the development of V2X technology, many researchers have contributed solutions to specific challenges of V2X communication protocol. Table 8.1 shows a comparison of V2X communication technologies. The Inter-Vehicle Hazard Warning (IVHW) system is one of the earliest studies to take the idea of improving vehicle safety based on communication [106]. The project is funded by the German Ministry of Education and Research and the French government. IVHW is a communication system in which warning messages are transmitted as broadcast messages in the frequency band of 869 MHz [84]. IVHW takes a local decision-making strategy. After the vehicle receives the message, it will do relevant checks to decide whether the warning message is relevant and should be shown to the driver. The majority of the research efforts have been made on the design of relevance check algorithms. However, as IVHW takes a broadcast mechanism to share the message, there can be a huge waste in both bandwidth and computing resources.

Table 8.1 The evolution of V2X communication technology

Research	Application scenario	Proposed solutions
IVHW	Safe driving	Warning message are transmitted as broadcast message; vehicle takes a local decision-making strategy.
FleetNet	Safe driving, internet protocol-based applications	Uses ad hoc networking to support multi-hop inter-vehicle communications; proposes a position-based forwarding mechanism.
CarTALK 2000	Cooperative driver assistance applications	Uses ad hoc communication network to support co-operative driver assistance applications; a spatial aware routing algorithm which takes some spatial information like underlying road topology into consideration to solve.
AKTIV	Safe driving	The latency of safety-related applications required to be less than 500 ms.
WILLWARN	Warning applications	Risk detection based on in-vehicle data; the warning message includes obstacles, road conditions, low visibility, and construction sites; a decentralized distribution algorithm to transmit the warning message to vehicles approaching the danger spot through V2V communication.
NoW	Mobility and internet applications	A hybrid forwarding scheme considering both network layer and application layer is developed; some security and scalability issues are discussed.
SAFESPOT	Safe driving	An integrated project that aims at using road-side infrastructure to improve driving safety; detects dangerous situations and shares the warning messages in real-time.
simTD	Traffic manipulation, safe driving, and internet-based applications	Real environment deployment of the whole intelligent transportation system; the system architecture includes three parts: ITS vehicle station, ITS road-side station, and ITS central station.

Compared with the broadcast message in IVHW, ad hoc networking can be a better solution to support multi-hop inter-vehicle communication [149]. FleetNet is another research project taking the idea of vehicle communication [198], and it is based on ad hoc networking. In addition, the FleetNet project also wants to provide a communication platform for some internet protocol-based applications. FleetNet is implemented based on IEEE 802.11 Wireless LAN system [139]. For vehicle-to-vehicle communication, if two vehicles are not directly connected wirelessly, it would need some other vehicles to forward the message for them. Designing the routing and forwarding protocol can be a major challenge. In order to meet the requirements for adaptability and scalability, FleetNet proposed a position-based forwarding mechanism. The idea is to choose the next hop to forward the message based on the geographical location of the vehicle.

CarTALK 2000 is also a project working on using ad hoc communication network to support co-operative driver assistance applications [438]. There can be a major challenge for ad hoc-based routing in vehicle-to-vehicle communication because the vehicle network topology is dynamic and the number of vehicles is frequently changing [363]. In order to solve the problem, a spatial aware routing algorithm is proposed in CarTALK 2000, which takes some spatial information like underlying road topology into consideration. Compared with FleetNet, CarTALK 2000 achieves better performance as it uses spatial information as additional input for routing algorithm. Another similar part of CarTALK 2000 and FleetNet is that they both are based on WLAN technology. AKTIV is another project which is the first one tried to apply cellular systems in some driving safety applications [80]. One of the reasons that FleetNet and CarTALK 2000 project built their system based on WLAN technology is that the latency of safety-related applications required is less than 500ms. However, with the assumption that an LTE communication system can be greatly further developed, cellular systems can be a good choice for sparse vehicle networking.

Meanwhile, some research projects have focused on warning applications based on vehicle-to-vehicle communication. Wireless Local Danger Warning (WILL-WARN) proposed a risk detection approach based on in-vehicle data. The warning message includes obstacles, road conditions, low visibility, and construction sites [473]. Unlike other projects focusing on the V2X technology itself, WILLWARN focuses on enabling V2X technology in some specific scenario such as the danger spot. Suppose some potential danger is detected in a specific location, but there is no vehicle within the communication range that supports the V2X communication technology to share the warning message [209]. To share warning messages, WILL-WARN proposed a decentralized distribution algorithm to transmit the warning message to vehicles approaching the danger spot through V2V communication. The project Network on Wheels (NoW) was one work that takes the idea of FleetNet to build vehicle communication based on 802.11 WLAN and ad hoc networking [140]. The goal of NoW is to set up a communication platform to support both mobility and internet applications. For example, a hybrid forwarding scheme considering both network layer and application layer is developed. Also, some security and scalability issues are discussed in NoW.

As the infrastructure also plays a very important part in V2X technology, several efforts focus on building safety applications based on the cooperation with infrastructure. SAFESPOT is an integrated project which is aimed at using road-side infrastructure to improve driving safety [524]. Through combining information from the on-vehicle sensors and infrastructure sensors, SAFESPOT detects dangerous situations and shares the warning messages in real-time. Also, the warning forecast can be improved from milliseconds level to seconds level, thus giving the driver more time to prepare and take action. Five applications are discussed in SAFESPOT, including hazard and incident warning, speed alert, road departure prevention, cooperative intersection collision prevention, and safety margin for assistance and emergency vehicles [53].

In 2007, a non-profit organization called the Car 2 Car Communication Consortium (C2C-CC) was set up to combine all solutions from different former projects to make a standard for V2X technology. Since 2010, the focus of work on V2X technology has moved from research topics to the real environment deployment of the whole Intelligent Transportation System (ITS). One of the most popular deploy projects is simTD [496], targeted on testing the V2X applications in a real metropolitan field. In simTD, all vehicles can connect with one another through Dedicated Short Range Communications (DSRC) technology which is based on IEEE 802.11p. Meanwhile, vehicles can also communicate with road-side infrastructure using IEEE 802.11p. The system architecture of simTD can be divided into three parts: ITS vehicle station, ITS road-side station, and ITS central station. Applications for testing in simTD include traffic situation monitoring, traffic flow information and navigation, traffic management, driving assistance, local danger alert, and internet-based applications.

Cellular Vehicle-to-Everything (C-V2X) is designed as a unified connectivity platform which provides low latency V2V and V2I communications [403]. It consists of two modes of communications. The first mode uses direct communication links between vehicles, infrastructure, and pedestrian. The second mode relies on network communication, which leverages cellular networks to enable vehicles to receive information from the internet. C-V2X further extends the communication range of the vehicle and it supports higher capacity of data for information transmission for vehicles.

8.1.2 Cooperative Autonomous Driving

Cooperative autonomous driving can be divided into two categories: one is cooperative sensing and the other is cooperative decision [213]. Cooperative sensing focuses on sharing sensing information between V2V and V2I. This data sharing can increase the sensing range of autonomous vehicles, making the system more robust. The cooperative decision enables a group of autonomous vehicles to cooperate and make decisions.

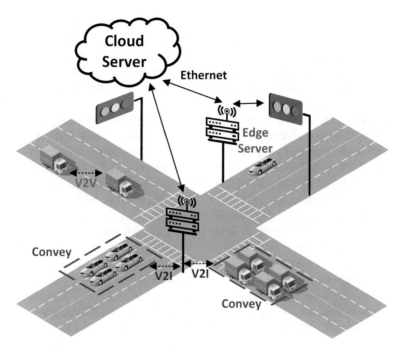

Fig. 8.1 V2X communications in crossroads

Some studies have focused on the exploration of applications for cooperative autonomous driving. In [213], four use cases including convoy driving, cooperative lane change, cooperative intersection management, and cooperative sensing are demonstrated. According to the design of AutoNet2030 [99], a convoy is formed of vehicles on multi-lanes into a group and the control of the whole group is decentralized. The safety and efficient control of the convoy requires high-frequency exchanges of each vehicle's dynamic data. As is shown in Fig 8.1, a road-side edge server and a cloud server are used to coordinate and manage the vehicles and convoys to go through crossroads safely. One convoy control algorithm in [344] only exchanges dynamic information of the nearby vehicle rather than for all the vehicles within a convoy. This design makes the algorithm easy to converge.

Cooperative lane change is designed to make vehicles or convoys to collaborate when changing lanes. Proper cooperative lane change cannot only avoid traffic accidents but it also reduces traffic congestion [260]. MOBIL [259] is a general model whose objective is to minimize overall braking induced by lane changes.

Cooperative intersection is also helpful for safe driving and traffic control. The World's Smartest Intersection in Detroit [353] focuses on safety and generates data that pinpoints areas where traffic-related fatalities and injuries can be reduced. Effective cooperative intersection management is based on coordination mechanism between vehicles to vehicles and vehicle to infrastructure.

Cooperative sensing increases the autonomous vehicle sensing range through V2X communication. Meanwhile, cooperative sensing also helps in cutting the cost of building autonomous driving. As vehicles can rely more on the sensors deployed on road-side infrastructure, the cost of on-vehicle sensors can be reduced. In the future, sensor information may become a service to the vehicle provided by the road-side infrastructure.

V2X networking infrastructure is also a very important aspect for cooperative autonomous driving. Heterogeneous Vehicular NETwork (HetVNET) [607] is an initial work on networking infrastructure to meet the communication requirements of the ITS. HetVNET integrates Long-Term Evolution (LTE) with DSRC [258], because relying on the single wireless access network cannot provide satisfactory services in dynamic circumstances. In [608], an improved protocol stack is proposed to support multiple application scenarios of autonomous driving in Heterogeneous Vehicular NETwork (HetVNET). In the protocol, the authors redefined the control messages in HetVNET to support autonomous driving.

Similarly, the Vehicular Delay-Tolerant Network (VDTN) [234] is an innovative communication architecture, which is designed for scenarios with long delays and sporadic connections. The idea is to allow messages to be forwarded in short range WiFi connections and reach the destination asynchronously. This property enables VDTN to support services and applications even when there is no end-to-end path in current VANET. Dias et al. [109] discuss several cooperation strategies for VDTN. The challenge for cooperation in VDTN is how to coordinate the vehicle node to share their constrained bandwidth, energy resources, and storage with one another. Furthermore, an incentive mechanism that rewards or punishes vehicles for the cooperative behavior is proposed.

In order to support seamless V2X communication, handover is also a very important topic for V2X networking infrastructure. Due to the dynamic changing of the networking topology and the relatively small range of the communication coverage, the handover mechanism in cellular network is no longer suitable for VANET. Based on proactive resource allocation techniques, Ghosh et al. [169] propose a new handover model for VANET. With the help of proactive handover, cooperative services can be migrated through Road-Side Units (RSUs) with the moving of the vehicle. Hence, proper designing of proactive handover and resource allocation are essential for developing reliable and efficient cooperative systems.

The development of edge computing in the automotive industry is also very inspiring. Automotive Edge Computing Consortium (AECC) is a group formed by automotive companies to promote edge computing technologies in future automobiles [24]. According to an AECC white paper from 2018, the service scenarios include intelligent driving, high-definition map, V2Cloud cruise assist, and some extended services like finance and insurance. In addition, the white paper discusses the service requirements in terms of data source, volume of data generated in vehicle, target data traffic rate, response time, and required availability.

8.1.3   Challenges

In order to guarantee the robustness and safety of autonomous driving systems, autonomous vehicles are typically equipped with numerous expensive sensors and computing systems, leading to extremely high costs and preventing ubiquitous deployment of autonomous vehicles. Hence, V2X is a viable solution in decreasing the costs of autonomous driving vehicles as V2X enables information sharing between vehicles and computation offloading to Road-Side Units. There are several challenges in achieving cooperative autonomous driving. Here, we discuss the challenges and our vision for application scenario of cooperative decision and cooperative sensing.

Cooperative Decision   The challenge of cooperative decisions is handling the dynamic changing topology with a short range coverage of V2X communications. The design of VDTN is a good hint to solve this challenge. Effective proactive handover and resource allocation can be a potential solution. Also, the coming 5G wireless communication [13] also provides a way to handle this challenge.

Cooperative Sensing   The main challenge of cooperative sensing is sharing the information from infrastructure sensors to autonomous vehicles in real-time, and the other challenge is to dynamically trade off the cost of infrastructure sensors and on-vehicle sensors. For the first challenge, the promising edge computing technology can be used to solve the problem [600], because edge computing enables the edge node (vehicle) and edge server (infrastructure) to conduct computation and compression to provide real-time performance. In addition, the trade-off of cost on infrastructure sensors and on-vehicle sensors will be determined by the automobile market. Both the government and companies will invest much money to support edge intelligence.

8.2   A Comparison of Communication Mechanisms in Vehicular Edge Computing

8.2.1   Introduction

Due to the safety and efficiency in fuel consumption, autonomous driving techniques have attracted huge attention from both academic and industry communities [317]. According to [174], the global autonomous driving market is expected grow up to $173.15B by 2030. However, there are still several challenges in the development and deployment of the autonomous driving system. The first challenge is on the cost. According to [177, 494], the cost of a level 4 autonomous driving vehicle can attain 300,000 dollars, in which the sensors and computing platform cost almost 200,000 dollars. The second challenge is the real-time requirements of the sensing, perception, and decision. According to [254], when the vehicle drives at 40 km per hour in urban areas and that autonomous functions should be effective every 1 m, the execution of each real-time task should be less than 100 ms. However, limited by the performance of perception algorithms and the stochastic runtime on the computing platform, how to guarantee the real-time requirement of the safety-critical system is still an open question. The third challenge is the reliability of cameras and LiDAR under severe weather conditions. The quality of images captured by cameras is affected by the lighting conditions [316], while the point clouds from LiDAR can be affected by noisy.

In order to address the above challenges, vehicle-to-everything (V2X)-based vehicle edge computing attracts more and more attention. V2X enables the vehicle to obtain real-time traffic information from the Road-Side Unit (RSU), which helps to decrease the cost of sensors on an autonomous driving vehicle. Meanwhile, there would be less sensing and perception tasks with real-time requirements and the reliability of cameras and LiDAR would not be an issue any more. Building intelligent traffic infrastructures with complex computation and communication resources becomes a consensus of the transportation community for the future transportation system.

For the state-of-the-art of vehicle edge computing (VEC), lots of works are focused on task offloading algorithms to optimize the usage of the V2X communications resources. In [132], a joint offloading decision and resource allocation model is designed to optimize the performance of task offloading when the computation requirement is unknown. Similarly, Tran et al. [525] propose a service caching and task offloading model to optimize the cost of the computation, while Chen and Hao [77] consider the offloading problems for the energy efficiency. However, all these works are based on simulations and the evaluation in real environment is missing. Besides, for the edge computing enabled application prototypes discussed in [314, 543], the communication mechanisms are limited to WiFi. Other potential communication mechanisms like LTE and DSRC are not touched. A general and comprehensive comparison of different types of V2X communication mechanisms for VEC applications is still missing.

In this chapter, we set up a VEC prototype, which supports three typical V2X communications including WiFi, Long-Term Evolution (LTE), and Dedicated Short Range Communication (DSRC). We set up dedicated end-to-end WiFi communication through a router, an LTE communication using Software-Defined Radio (SDR) device USRP B210 boards with wireless antennas, and a DSRC communication using commercial products with On-Board Unit (OBU) and Road-Side Unit (RSU). On top of the prototype, several applications including *Robot Operating System (ROS)*, *Socket*, and *Ping* messages are implemented to evaluate the performance of the communication as well as their impacts to the computation [424, 533]. The evaluation covers the round-trip latency, system utilization, and power dissipation. We summarize several observations toward the usability, availability, and efficiency of the V2X communications to the VEC applications. The contributions of this work are as follows:

- To the best of our knowledge, our work is the first comparison of communication mechanisms of LTE, WiFi, and DSRC using real VEC applications.
- An end-to-end communication prototype is built, which can support LTE, DSRC, and WiFi. On top of the prototype, *ROS*, *Socket*, and *Ping* messages are implemented.
- We evaluated the communication prototype in latency, power dissipation, and system utilization and get three observations for the real deployment of VEC applications.

8.2.2 *Background*

Considering a typical VEC scenario shown in Fig. 8.2, the cloud, base station (BS), RSU, and vehicles are connected through a three-tier communication network. The bottom tier is for vehicles, where vehicles can share information through DSRC, LTE, and WiFi. The middle tier includes the BS and RSU. Vehicles communicate with the RSU through LTE, DSRC, and LTE, while they can also communicate with the BS through LTE. The top tier is the cloud data center. Both BS and RSU are connected to the cloud through Ethernet or optical fiber connections.

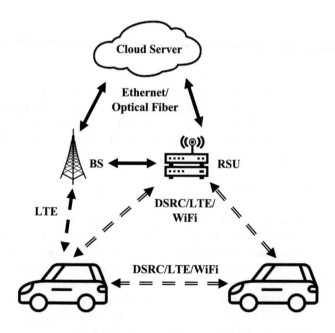

Fig. 8.2 A typical example of vehicular edge computing

Table 8.2 A comparison of V2X communications

Communications	Bandwidth	Max range
WiFi (802.11ac)	2.4 GHz/5 GHz	Indoor: 46 m Outdoor: 92 m
4G/LTE	67 MHz	–
DSRC	75 MHz	1000 m (120 mph)

Table 8.2 shows a brief comparison of LTE, DSRC, and WiFi in terms of frequency and max range. In practical, DSRC is designed for sharing information between vehicles and traffic infrastructure. There are fifteen message types that are defined in the SAE J2735 standard [492], which covers information like the vehicle's position, map information, emergence warning, etc. [258]. Limited by the available bandwidth, DSRC messages have small size and low frequency. However, DSRC provides reliable communication even when the vehicle is driving 120 miles per hour. In contrast, WiFi and LTE provide more bandwidth but perform poor in mobile environment. The coverage is another concern when implementing the V2X application. Due to the mobile cellular network technology, LTE is the best in terms of coverage, while DSRC also achieves almost 1000 meters' coverage with the speed at 120 mph. The coverage of WiFi is different for indoor and outdoor scenarios. For 802.11ac, it achieves 46 meters for indoor scenario while 92 meters for outdoor.

8.2.3 Methodology

In order to compare the performance of different V2X communication mechanisms and evaluate their impact to the computation, we build an end-to-end prototype which supports WiFi, LTE, and DSRC. In this section, first we discuss their setups, and then we discuss the communication messages defined for the comparison.

8.2.3.1 Hardware Description

For WiFi-based communication, a TPlink router is used to set up hotspots. The router supports two frequency bands: 2.4 and 5 GHz. For LTE-based communication, two Ettus Research USRP B210 boards with two VERT2450 antennas are used with *uhd* software driver to set up the network. For DSRC-based communication, we use moKar DSRC devices with two DSRC antennas in the prototype.

All the communications antennas/devices are connected to a host machine to handle the communication and application development. For the on-board computing boards, we use Intel Fog Reference and Nvidia Jetson TX2. The Intel Fog Reference has 8 Intel Xeon(R) CPU with 3.60 GHz frequency. And it has 32 GB memory. Jetson TX2 is a typical mobile computing board with GPU installed [548]. For RSU, a cluster of Intel Fog Reference boards are connected to provide computation as well as communication resources to vehicles [312]. NVIDIA® Jetson TX2 is a power-efficient embedded AI computing platform, which has dual-core Denver and quad-core ARM® Cortex-A57 CPU equipped with 8 GB DDR4 memory and 32 GB eMMC. The GPU on Jetson TX2 is powered by NVIDIA Pascal™architecture with 256 CUDA cores.

Four types of messages are leveraged to transmit in the end-to-end bidirectional prototype. We choose three ROS-based messages and one Internet Control Message Protocol (ICMP)-based message: *BSM*, *image1*, *image2*, and *Ping*. Basic Safety Message (BSM) is one of the fifteen predefined messages to share the vehicle's position, speed, and direction, with timestamp. We choose *BSM* and implement the message sharing through ROS. The size of *BSM* message is 1416 bytes. In addition, two image *ROS* messages are defined and they include the header which has the timestamp, size, an array to store image pixels, etc. We choose two images: one is 91 kB and another is 401 kB. *Ping* is another message we used for the evaluation of latency and power dissipation. The size of *Ping* message is 64 bytes. Since DSRC-based communication does not generate a virtual IP address and it does not support transmitting images, only the predefined BSM message is implemented. For WiFi and LTE, all four messages are implemented and evaluated.

8.2.3.2   WiFi-Based Communication

As one of the most popular wireless communications for large volume of data transmission, WiFi is a potential communication method for VEC applications. Figure 8.3 shows the communication framework implemented on top of WiFi. When OBU and RSU connected to the router, unique IP addresses are assigned and they are able to *Ping* each other within the local area network. In this chapter, we use 5 GHz WiFi connection.

On top of the WiFi connection, we implement an *ROS-based* communication for data sharing. ROS is a communication middleware designed for robots as well as

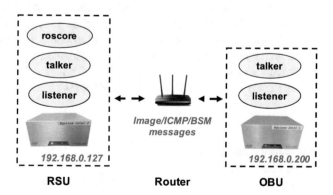

Fig. 8.3 WiFi-based communication

connected and autonomous driving vehicles [443]. In ROS, a master is launched to manage all the ROS nodes, topics, services, and actions. *roscore* is used to launch the ROS master in RSU. Two ROS nodes are launched in both RSU and OBU: talker and listener. They are defined to send ROS messages between RSU and OBU through inter-process communication (IPC).

8.2.3.3   LTE-Based Communication

Compared with WiFi, LTE has good strength on the coverage of the communication and its performance in mobile environment. Driven by the development of Software-Defined Radio (SDR), the programmability of the wireless communication devices improves a lot and it becomes easier to set up a dedicated LTE communication testbed using SDR boards with open-sourced software like *OpenAirInterface* and *srsLTE* [175, 385].

In our prototype, the LTE communication is set up by using two Ettus Research USRP B210 boards with VERT2450 antennas [533]. USRP B210 boards are connected to the host machine through USB 3.0 interface. For the software, *uhd* is built as the USRP driver and *srsLTE* is installed to build the whole LTE network. *srsLTE* is a free and open-source 4G/LTE software suite which includes core network (EPC), base station (eNB), and user equipment (UE) [175]. All the modems can be implemented as an application running in a standard Linux-based machine.

As is shown in Fig. 8.4, EPC and eNB are implemented on the RSU because they need to provide LTE connections to all nearby vehicles. UE is implemented on the OBU for registration and connection to the eNB. To begin with, EPC with eNB is launched to broadcast signals at a specific frequency. After UE is launched on another machine, it will search for the area to find the cellular tower and try to set up connection. When the UE is registered in EPC and connected with the eNB, a unique IP address will be assigned to the UE. For example, a virtual network interface is implemented by eNB on the RSU (172.16.0.1) and it assigns 172.16.0.2 to the UE. With unique IP addresses in the local area network, RSU and OBU can ping each other and the networking programming interface can be used for application development. Similar to WiFi, the data sharing application is also implemented based on ROS. ROS master with talker and listener ROS nodes is implemented on the RSU, while talker and listener ROS nodes are implemented on the OBU. We use the default frequency of *srsLTE* called EARFCN 3400 with single VERT 2450 antenna, which represents the download link with 2685 MHz and upload link with 2565 MHz.

8.2.3.4   DSRC-Based Communication

DSRC-based communication enables RSU and OBU to share short messages such as safety warnings and traffic information with the communication frequency of 5.9 GHz band. Our prototype of DSRC-based communication is shown in Fig. 8.5.

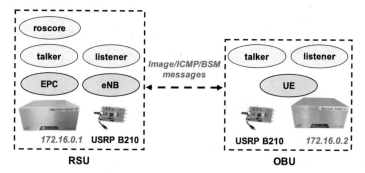

Fig. 8.4 LTE-based communication

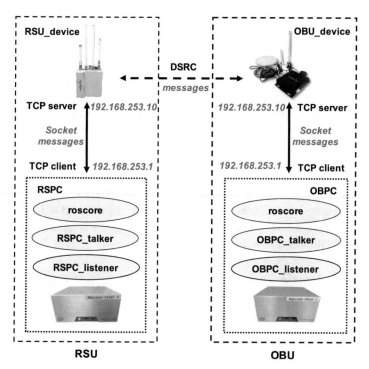

Fig. 8.5 DSRC-based communication

RSU and OBU can share messages with each other, while message sent from RSPC and OBPC can go through RSU_device and OBU_device. Host machines (RSPC and OBPC) and DSRC devices (RSU_device and OBU_device) are within the same LAN. DSRC communication is used between RSU_device and OBU_device.

For both of RSPC and OBPC, two ROS nodes named *talker* and *listener* are created. *talker* reports information of vessel, such as latitude, longitude, heading, and speed at frequency of 250 Hz, while listener subscribes to it with the same frequency. TCP Socket-based communication is set up between host machines and DSRC devices. The RSU_device and OBU_device communicate with each other through DSRC to share BSM messages at 250 Hz.

8.2.4   Comparison and Observation

In order to evaluate the performance of the end-to-end communication prototype, we conducted experiments and evaluated in three aspects: end-to-end latency, power dissipation, and system utilization.

8.2.4.1   End-to-End Latency

End-to-end latency is one of the most important metrics to evaluate the performance of applications which heavily relies on the communication. In this part, latency is measured in two directions: from RSU to OBU and from OBU to RSU. Since time difference affects the latency results, we set up Network Time Protocol (NTP)-based time synchronization and use *clockdiff* in Linux to measure the clock difference when transmitting messages. Then the clock difference is applied to calculate the end-to-end latency. The results of the average end-to-end delay are shown in Tables 8.3 and 8.4. From RSU to OBU, LTE performs the best for *BSM* and *Ping*, while WiFi is better for two image messages. From OBU to RSU, DSRC is the best for *BSM* message. For LTE-based communication, we notice that LTE performs better than WiFi for *BSM* and *Ping* but worse for *image*. In addition, DSRC shows stable performance in transmitting small messages.

Table 8.3 The end-to-end latency from RSU to OBU

RSU -> OBU (ms)	LTE	WiFi	DSRC
BSM	**3.12**	67.96	8.46
image1	6174.32	**1733.90**	–
image2	20220.85	**1741.07**	–
Ping	**25.02**	50.49	–

Table 8.4 The end-to-end latency from OBU to RSU

OBU -> RSU (ms)	LTE	WiFi	DSRC
BSM	51.02	14.90	**9.84**
image1	6012.8	**1301.38**	–
image2	21876.4	**1293.25**	–
Ping	**28.22**	67.52	–

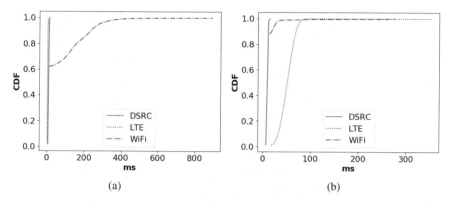

Fig. 8.6 The CDF of *BSM* message end-to-end latency in (**a**) RSU to OBU and (**b**) OBU to RSU

Table 8.5 The latency of transmitting BSM message with different speed

Latency (ms)	LTE	WiFi	DSRC
0 m/s	**3.12**	15.06	8.88
3 m/s	657.28	12.99	**9.12**
5 m/s	1071.52	16.18	**9.21**

Besides the average end-to-end delay, we also draw the CDF of the end-to-end delay in two directions for transmitting *BSM* message, which are shown in Fig. 8.6. We can find that WiFi shows large variance for RSU to OBU (over 300 ms), while LTE shows large variance for OBU to RSU (almost 100 ms). We think one reason for the variation of LTE is the bandwidth difference between downloading and uploading. The variance of DSRC is always the lowest, which means the latency of DSRC is very stable.

In order to measure the latency in mobile environment, HydraOne is also used in the experiments to communicate with the RSU [548]. We choose *BSM* message and measure the latency from RSU to OBU with LTE, WiFi, and DSRC. From the results shown in Table 8.5, DSRC shows the lowest difference when the speed of HydraOne changes, while LTE's latency increased by over 300 times when the speed increases from 0 to 5 m/s. For WiFi-based communication, negligible difference is observed when the speed changes. From the discussion above, we can generalize our first observation:

Observation 1 *DSRC shows fast and reliable performance for transmitting pre-defined messages. WiFi has the lowest latency when transmitting images, but its variance for BSM and ping is large. LTE performs well for small messages like BSM and ping but bad for large messages like images. LTE and DSRC have stable performance under mobile environment, while LTE has the most performance degradation. A communication mechanism that can provide sufficient and stable bandwidth is still missing.*

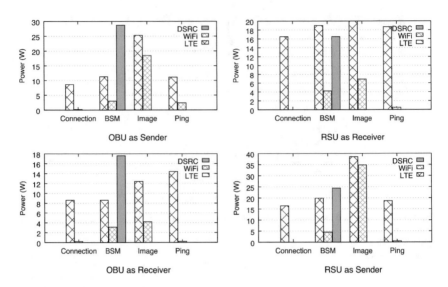

Fig. 8.7 Power dissipation

8.2.4.2   Power Dissipation

For the measurement of power dissipation, Watt's Up Pro is used to record the hardware power dissipation and *rapl* tool from Intel is used to measure the processor power dissipation [210]. Since both OBU and RSU can be sender and receiver, we measure the power dissipation for communication connection and message transmission in four cases, which are shown in Fig. 8.7.

From the results, we can find that WiFi is always more energy efficient than LTE and DSRC. For the transmission of *BSM* message, DSRC has the most energy consumption for most cases and they are all larger than 16 W. In addition to the data transmission, communication setup/connection also consumes a lot of energy. The connection of LTE consumes 8.6 W for OBU and 16.5 W for RSU, which is the power overhead of setting up EPC and eNB on RSU and UE on the OBU. Besides, transmitting image messages consumes more energy than others. On sender's side, the power dissipation for image sending is larger than 18.5W. This fact leads to the following observation:

Observation 2 *There is non-negligible power dissipation for communication connections for LTE and DSRC. Since sending messages also consumes lot of energy, how to make the V2X communication more energy efficient is still an open question.*

Table 8.6 The memory footprint of RSU as sender and receiver

RSU (kb)	msg	LTE	WiFi	DSRC
Sender	*BSM*	49960	50084	**19700**
	image	264872	**264700**	–
Receiver	*BSM*	50364	264872	**1588**
	image	**120808**	125196	–

Table 8.7 The CPU utilization of RSU as sender and receiver

RSU (%)	msg	LTE	WiFi	DSRC
Sender	*BSM*	1	2.63	**1.036**
	image	120.53	**100.6**	–
Receiver	*BSM*	1.18	120.53	**0.1**
	image	1	9.03	–

Table 8.8 The memory footprint of OBU as sender and receiver

OBU (kb)	msg	LTE	WiFi	DSRC
Sender	*BSM*	51832	51968	**20100**
	image	**165996**	312400	–
Receiver	*BSM*	53408	51924	**10224**
	image	**126380**	126688	–

Table 8.9 The CPU utilization of OBU as sender and receiver

OBU (%)	msg	LTE	WiFi	DSRC
Sender	*BSM*	1.2	1.11	**1.08**
	image	**91.44**	99.76	–
Receiver	*BSM*	1.25	1.1	**0.124**
	image	**1.36**	1.78	–

8.2.4.3 System Utilization

For the measurement of system utilization, we use *top* to get the memory footprint and the CPU utilization. The results are shown in Tables 8.6, 8.7, 8.8, and 8.9. It can be found that DSRC has the lowest memory footprint and CPU utilization for *BSM* message, while LTE has less memory footprint and CPU utilization when RSU is the receiver. For WiFi, the memory footprint and the CPU utilization are higher than others except when RSU is sending image message.

In addition to the system utilization for sending messages, there is also some overhead in setting up the communications. On RSU, the extra overhead includes launching ROS master node through *roscore* and launching the EPC with eNB for LTE communications. On OBU, UE needs to be launched for LTE connection. The results for the above overhead are shown in Table 8.10. We can find that ROS master does not consume a lot of memory footprint, but eNB and UE consume large memory footprint and CPU resources. This fact leads to the following observation:

Observation 3 *Sending process consumes more memory footprint and CPU resources than receiving. The usage of srsLTE as well as ROS introduces non-negligible system overhead.*

Table 8.10 The overhead of WiFi and LTE communications

Metrics	RSU			OBU
	Roscore LTE	Roscore WiFi	eNB	UE
Memory (kb)	50364	55400	**1599532**	**1488000**
CPU (%)	2.22	1	**27.73**	**23.56**

8.2.5   Summary

CAVs have attracted massive attention from both the academic and automotive communities. The driving system plays a crucial role in understanding the traffic environment and making decisions for CAVs. However, the limited system-level reliability and the missing integration with vehicle communications and controls narrow the testing of CAVs to constraint scenarios. This chapter presents our vision of a future driving system for CAVs, called 4*C*, which provides a unified framework for communication, computation, and control co-design. With two case studies, we present the usability of the 4*C* framework. Finally, we discuss the challenges in the 4*C* framework.

Chapter 9
Challenges and Open Problems

9.1 Artificial Intelligence for Autonomous Driving

Most of the AV's services (e.g., environmental perception and path planning) are carried out by artificial intelligence-based approaches. As the focus of the automotive industry gradually shifts to series production, the main challenge is how to apply machine learning algorithms to mass-produced AVs for real-world applications. Here, we list three main challenges in artificial intelligence (AI) for AVs.

9.1.1 Standardization of Safety Issue

One of the main challenges is that machine learning algorithms are unstable in terms of performance. For example, even a small change to camera images (such as cropping and variations in lighting conditions) may cause the ADAS system to fail in object detection and segmentation [168, 401, 571]. However, the automotive safety standard of ISO 26262 [216] was defined without taking deep learning into consideration because the ISO 26262 was published before the boom of AI, leading to the absence of proper ways to standardize the safety issue when incorporating AI for AVs [428].

9.1.2 Infeasibility of Scalable Training

To achieve high performance, machine learning models used on AVs need to be trained on representative datasets under all application scenarios, which bring challenges in training time-sensitive models based on petabytes of data. In this case,

W. Shi, L. Liu, *Computing Systems for Autonomous Driving*,
https://doi.org/10.1007/978-3-030-81564-6_9

collaborative training [91], model compression technologies [95, 103, 104, 194, 329, 465], and lightweight machine learning algorithms [87, 222, 230] were proposed in recent years. Besides, getting accurate annotations of every pedestrian, vehicle, lane, and other objects is necessary for the model training using supervised learning approaches, which becomes a significant bottleneck [586].

9.1.3 Infeasibility of Complete Testing

It is infeasible to test machine learning models used on AVs thoroughly. One reason is that machine learning learns from large amounts of data and stores the model in a complex set of weighted feature combinations, which is not intuitive or difficult to conduct thorough testing [274]. In addition, previous work pointed out that, to verify the catastrophic failure rate, around 10^9 h (billion hours) of vehicle operation test should be carried out [63] and the test needs to be repeated many times to achieve statistical significance [428].

9.2 Multi-sensors Data Synchronization

Data on the autonomous driving vehicle has various sources: its sensors, other vehicle sensors, RSU, and even social media. One big challenge to handle a variety of data sources is how to synchronize them.

For example, a camera usually produces 30–60 frames per second, while LiDAR's point cloud data frequency is 10 HZ. For applications like 3D object detection, which requires camera frames and point cloud at the same time, should the storage system do synchronization beforehand or let the application developer do it? This issue becomes more challenging, considering that the timestamp's accuracy from different sensors falls into different granularities. For example, considering the vehicles that use Network Time Protocol (NTP) for time synchronization, the timestamp difference can be as long as 100 ms [181, 357]. For some sensors with a built-in GNSS antenna, the time accuracy goes to the nanosecond level. In contrast, other sensors get a timestamp from the host machine's system time when accuracy is at the millisecond level. Since the accuracy of time synchronization is expected to affect the vehicle control's safety, handling the sensor data with different frequency and timestamp accuracy is still an open question.

9.3 Failure Detection and Diagnostics

Today's AVs are equipped with multiple sensors, including LiDARs, radars, and GPS [574]. Although we can take advantage of these sensors in terms of providing a robust and complete description of the surrounding area, some open problems related to the failure detection are waiting to be solved. Here, we list and discuss four failure detection challenges: (1) *Definition of sensor failure:* there is no standard, agreed-upon universal definition or standards to define the scenario of sensor failures [453]. However, we must propose and categorize the standard of sensor failures to support failure detection by applying proper methods. (2) *Sensor failure:* more importantly, there is no comprehensive and reliable study on sensor failure detection, which is extremely dangerous since most of the self-driving applications are relying on the data produced by these sensors [397]. If some sensors encountered a failure, collisions and environmental catastrophes might happen. (3) *Sensor data failure:* in the real application scenario, even when the sensors themselves are working correctly, the generated data may still not reflect the actual scenario and report the wrong information to people [520]. For instance, the camera is blocked by unknown objects such as leaves or mud, or the radar deviates from its original fixed position due to wind force. In this context, sensor data failure detection is very challenging. (4) *Algorithm failure:* in challenging scenarios with severe occlusion and extreme lighting conditions, such as night, rainy days, and snowy days, deploying and executing state-of-the-art algorithms cannot guarantee output the ideal results [509]. For example, lane markings usually fail to be detected at night by algorithms that find it difficult to explicitly utilize prior information like rigidity and smoothness of lanes [506]. However, humans can easily infer their positions and fill in the occluded part of the context. Therefore, how to develop advanced algorithms to further improve detection accuracy is still a big challenge.

For a complex system with rich sensors and hardware devices, failures could happen everywhere. How to tackle the failure and diagnosing the issue becomes a big issue. One example is the diagnose of lane controller systems from Google [285]. The idea is to determine the root cause of malfunctions based on comparing the actual steering corrections applied to those predicted by the virtual dynamics module.

9.4 How to Deal with Normal–Abnormal?

Normal–abnormal represents normal scenarios in daily life that are abnormal in the autonomous driving dataset. Typically, there are three cases of normal–abnormal: adverse weather, emergency maneuvers, and work zones.

9.4.1 Adverse Weather

One of the most critical issues in the development of AVs is the poor performance under adverse weather conditions, such as rain, snow, fog, and hail, because the equipped sensors (e.g., LiDAR, radar, camera, and GPS) might be significantly affected by the extreme weather. The work of [591] characterized the effect of rainfall on millimeter-wave (mm-wave) radar and proved that under heavy rainfall conditions, the detection range of millimeter-wave radar can be reduced by as much as 45%. Filgueira et al. [142] pointed out that as the rain intensity increases, the detected LiDAR intensity will attenuate. At the same time, Bernardin et al. [42] proposed a methodology to quantitatively estimate the loss of visual performance due to rainfall. Most importantly, experimental results show that, compared to training in narrow cases and scenarios, using various data sets to train object detection networks may not necessarily improve the performance of these networks. [212]. However, there is currently no research to provide a systematic and unified method to reduce the impact of weather on various sensors used in AVs. Therefore, there is an urgent need for novel deep learning networks that have sufficient capabilities to cope with safe autonomous driving under severe weather conditions.

9.4.2 Emergency Maneuvers

In emergency situations, such as a road collapse, braking failure, a tire blowout, or suddenly seeing a previously "invisible" pedestrian, the maneuvering of the AVs may need to reach its operating limit to avoid collisions. However, these collision avoidance actions usually conflict with stabilization actions aimed at preventing the vehicle from losing control, and in the end, they may cause collision accidents. In this context, some research has been done to guarantee safe driving for AVs in emergent situations. For example, Hilgert et al. proposed a path planning method for emergency maneuvers based on elastic bands [207]. The paper [153] is proposed to determine the minimum distance at which obstacles cannot be avoided at a given speed. Guo et al. [190] discussed dynamic control design for automated driving, with particular emphasis on coordinated steering and braking control in emergency avoidance. Nevertheless, how an autonomous vehicle safely responds to different classes of emergencies with on-board sensors is still an open problem.

9.4.3 Work Zone

Work zone recognition is another challenge for an autonomous driving system to overcome. For most drivers, the work zone means congestion and delay of the driving plan. Many projects have been launched to reduce and eliminate work zone injuries and deaths for construction workers and motorists. "Workzonesafety.org" summarizes recent years of work zone crashes and supplies training programs to increase public awareness of the importance of work zone safety. Seo [479] proposed a machine learning-based method to improve the recognition of work zone signs. Developers from Kratos Defense & Security Solutions [35] present an autonomous truck which safely passes a work zone. Their system relied on V2V communications to connect the self-driving vehicle with a leader vehicle. The self-driving vehicle accepted navigation data from the leader vehicle to travel along its route while keeping a predefined distance. Until now, the work zone is still a threat to drivers and workers' safety but has not attracted too much attention to autonomous driving researchers. There are still significant gaps in this research field, waiting for researchers to explore and tackle critical problems.

9.5 Cyberattack Protection

Attacks and defenses are always opposites, and absolute security does not exist. The emerging CAVs face many security challenges, such as reply attacks to simulate a vehicle's electronic key and spoof attacks to make vehicle detour [242, 440]. With the integration of new sensors, devices, technologies, infrastructures, and applications, the attack surface of CAVs is further expanded.

Many attacks focus on one part of the CAVs system and could be protected by the method of fusing several other views. For example, a cheated roadblock detected by radars could be corrected by camera data. Thus, how to build such a system to protect CAVs, systematically, is the first challenge for the CAVs system. The protection system is expected to detect potential attacks, evaluate the system security status, and recover from attacks.

Besides, some novel attack methods should be attended. Recently, some attacks have been proposed to trick these algorithms [68]. For example, a photo instead of a human to pass the face recognition or a note-sized photo posted on the forehead makes machine learning algorithms fail to detect faces [269]. Thus, how to defend the attacks on machine learning algorithms is a challenge for CAVs systems.

Furthermore, some new technologies could be used to enhance the security of the CAVs system. With the development of quantum computing technology, the existing cryptography standards cannot ensure protected data, communication, and systems. Thus, designing post-quantum cryptography [43] and architecture is a promising topic for CAVs and infrastructure in ITS.

Also, we noticed that the hardware-assistant trusted execution environment [387] could improve the system security, which provides an isolated and trusted execution environment (TEE) for applications. However, it has limited physical memory size, and execution performance will drop sharply as the total memory usage increases. Therefore, how to split the system components and make critical parts in the TEE with high security is still a challenge in design and implementation.

9.6   Vehicle Operating System

The vehicle operating system is expected to abstract the hardware resources for higher layer middleware and autonomous driving applications. In the vehicle operating system development, one of the biggest challenges is the compatibility with the vehicle's embedded system. Take Autoware as an example: although it is a full-stack solution for the vehicle operating system that provides a rich set of self-driving modules composed of sensing, computing, and actuation capabilities, the usage of it is still limited to several commercial vehicles with a small set of supportable sensors [255]. On a modern automobile, as many as 70 ECUs are installed for various subsystems, and they are communicated via CAN bus. For the sake of system security and commercial interests, most of the vehicles' CAN protocol is not open-sourced, which is the main obstacle for developing a unified vehicle operating system.

AUTOSAR is a standardization initiative of leading automotive manufacturers and suppliers founded in the autumn of 2003 [19]. AUTOSAR is promising in narrowing the gap for developing an open-source vehicle operating system. However, most automobile companies are relatively conservative to open source their vehicle operating systems, restricting the availability of AUTOSAR to the general research and education community. There is still a strong demand for a robust, open-source vehicle operating system for AVs.

9.7   Energy Efficiency

With rich sensors and powerful computing devices implemented on the vehicle, energy consumption becomes a big issue. Take the NVIDIA Drive PX Pegasus as an example: it consumes 320 INT8 TOPS of AI computational power with a 500 watts budget. If we added external devices like sensors, communication antennas, storage, battery, etc., the total energy consumption would be larger than 1000 W [350]. Besides, if a duplicate system is installed for the autonomous driving applications' reliability, the total power dissipation could go up to almost 2000 W.

How to handle such a tremendous amount of power dissipation is not only a problem for the battery management system but also a problem for the heat dissipation system. What makes this issue more severe is the size limitation and auto-grid requirements from the vehicle's perspective. How to make the computing system of the autonomous driving vehicle become energy efficient is still an open challenge. E2M tackles this problem by proposing as an energy-efficient middleware for the management and scheduling deep learning applications to save energy for the computing device [309]. However, according to the profiling results, most of the energy is consumed by vehicles' motors. Energy-efficient autonomous driving requires the co-design in battery cells, energy management systems, and autonomous vehicle computing systems.

9.8 Cost

In the United States, the average cost to build a traditional non-luxury vehicle is roughly $30,000, and for an AV, the total cost is around $250,000 [291]. AVs need an abundance of hardware equipment to support their normal functions. Additional hardware equipments required for AVs include but are not limited to the communication device, computing equipment, drive-by-wire system, extra power supply, various sensors, cameras, LiDAR, and radar. The approximate prices of each equipment are shown in Table 9.1. In addition, to ensure AV's reliability and safety, a backup of these hardware devices may be necessary [476]. For example, if the main battery fails, the vehicle should have a backup power source to support computing systems to move the vehicle.

The cost of building an autonomous vehicle is already very high, not to mention the maintenance cost of an AV, e.g., diagnostics and repair. High maintenance costs lead to declining consumer demand and undesirable profitability for the vehicle manufacturers. Companies like Ford and GM have already cut their low-profit production lines to save costs [188, 328].

Indeed, the cost of computing systems for AVs currently in the research and development stage is very high. However, we hope that with the maturity of the technologies and the emergence of some alternative solutions, the price will ultimately drop to a level that individuals can afford. Take battery packs of electric vehicles (EVs) as an example: when the first mass-market EVs were introduced in 2010, their battery packs were estimated at $1000 USD per kilowatt-hour (kWh). However, Tesla's Model 3 battery pack costs $190 per kilowatt-hour, and General Motors' 2017 Chevrolet Bolt battery pack is estimated to cost $205 per kilowatt-hour. In 6 years, the price per kilowatt-hour has dropped by more than 70% [92].

Table 9.1 Hardware equipment of autonomous vehicles

Hardware equipment		Unit price	
Communication device		DSRC	$17,600
		LTE/5G/C-V2X	$177
Computation device		Consists of a PC, a hard drive, a graphics card, and other devices	$5000
Vehicle electronic Control unit (ECU)		Each vehicle has 50 to 100 ECUs	$320 each
Power supply		$24,300	
Sensor	Camera	$6000	
	LiDAR	$8,5000 (installed on the top, 360-degree field of view) + $8000 * 4 (4 small ones installed in 4 corners)	
	Radar	$500 each * 6	
	Other	E.g. inertial measurement unit, $4000	

Data listed as of August 2020

Also, Waymo claims to have successfully reduced the experimental version of high-end LiDAR to approximately $7500. Besides, Tesla, which uses only radar instead of LiDAR, says its autonomous vehicle equipment is around $8000 [291]. In addition to the reduction of hardware costs, we believe that the optimization of computing software in an AV can also help reduce the cost to a great extent.

9.9   How to Benefit from Smart Infrastructure?

Smart infrastructure combines sensors, computing platforms, and communication devices with the physical traffic infrastructure [57]. It is expected to enable the AVs to achieve more efficient and reliable perception and decision-making. Typically, AVs could benefit from smart infrastructure in three aspects: (1) *Service provider:* It is struggling for an AV to find a parking space in the parking lot. By deploying sensors like RFID on the smart infrastructure, parking services can be handled quickly [399]. As the infrastructure becomes a provider for parking service, it is possible to schedule service requests to achieve the maximum usage. Meanwhile, AVs can reduce the time and computation for searching services. (2) *Traffic information sharing:* Traffic information is essential to safe driving. Lack of traffic information causes traffic congestion or even accidents. Road-Side Units (RSUs) are implemented to provide traffic information to passing vehicles through V2X communications. Besides, RSUs are also used to surveil road situations using various on-board sensors like cameras and LiDARs [5]. The collected data is used for various tasks, including weather warning, map updating, road events detection, and making up blind spots of AVs. (3) *Task offloading:* Various algorithms are running on vehicles for safe driving. Handling all workloads in real-time requires a tremendous amount of computation and power, infeasible on a battery-powered vehicle [304]. Therefore, offloading heavy computation workloads to the infrastructure is proposed to accelerate the computation and save energy. However, to perform feasible offloading, the offloading framework must offload computations to the infrastructure while ensuring timing predictability [117]. Therefore, how to schedule the order of offloading workloads is still a challenge to benefit from the smart infrastructure.

9.10   Experimental Platform

The deployment of autonomous driving algorithms or prototypes requires complex tests and evaluations in a real environment, which makes the experimental platform become one of the fundamental parts of conducting research and development. However, building and maintaining an autonomous driving vehicle are enormous: the cost of a real autonomous driving vehicle could attain $250,000; maintaining the vehicle requires parking, insurance, and auto-maintenance. Let alone the laws and regulations to consider for field testing.

Given these limitations and problems, lots of autonomous driving simulators and open-source prototypes are proposed for research and development purposes. dSPACE provides an end-to-end simulation environment for sensor data processing and scenario-based testing with RTMaps and VEOS [126]. The automated driving toolbox is Mathwork's software, which provides algorithms and tools for designing, simulating, and testing ADAS and autonomous driving systems [26]. AVL Drive-Cube is a hardware-in-the-loop driving simulator designed for real vehicles with simulated environments [27]. In addition to these commercialized products, there are also open-source projects like CARLA and Gezabo for urban driving or robotics simulations [119, 267].

Another promising direction is to develop affordable research and development of autonomous driving platforms. Several experiment platforms are quite successful for indoor or low-speed scenarios. HydraOne is an open-source experimental platform for indoor autonomous driving, and it provides full-stack programmability for autonomous driving algorithm developers and system developers [548]. DragonFly is another example that supports self-driving with a speed of fewer than 40 miles per hour and a price of less than $40,000 [411].

9.11   Human Driver Interaction

According to NHTSA data collected from all 50 states and the District of Columbia, 37,461 lives were lost on U.S. roads in 2016, and 94% of crashes were associated with "a human choice or error" [439]. Although autonomous driving is proposed to replace human drivers with computers/machines for safety purposes, human driving vehicles will never disappear. How to enable computers/machines in AVs to interact with a human driver becomes a big challenge [455].

Compared with a human driver, machines are generally more suited for tasks like vehicle control and multi-sensor data processing. In contrast, the human driver maintains an advantage in perception and sensing the environment [471]. One of the fundamental reasons is that the machine cannot think like a human. Current machine learning-based approaches cannot handle situations that are not captured in the training dataset. For example, in driving automation from SAE, one of the critical differences between level 2 and level 3/4/5 is whether the vehicle can make decisions like overtaking or lane changing by itself [317]. In some instances, interacting with other human drivers becomes a big challenge because human drivers can make mistakes or violate traffic rules.

Many works focus on getting a more accurate speed and control predictions of the surrounding vehicles to handle the machine–human interaction [164, 226]. Deep reinforcement learning shows promising performance in complex scenarios requiring interaction with other vehicles [299, 456]. However, they are either simulation-based or demonstration in limited scenarios. Another promising direction to tackle machine–human interaction is through V2X communications. Compared with predicting other vehicles' behavior, it is more accurate to communicate safety information [102].

9.12 Physical Worlds Coupling

Autonomous driving is a typical cyber-physical system [179], where the computing systems and the physical world have to work closely and smoothly. With a human driver, the feeling of a driver is easily coupled with the vehicle control actions. For example, if the driver does not like the abrupt stop, he or she can step on the brake gradually. In autonomous driving, the control algorithm will determine the speed of braking and accelerating. We envision that different human feelings, coupled with complex traffic environment, bring an unprecedented challenge to the vehicle control in autonomous driving. Take the turning left as an example: how fast should the drive-by-wire system turn 90°? An ideal vehicle control algorithm of turning left should consider many factors, such as the friction of road surface, vehicle's current speed, weather conditions, and the movement range, as well as human comfortableness, if possible. Cross-layer design and optimization among perception, control, vehicle dynamics, and drive-by-wire systems might be a promising direction [331].

References

1. Abadi, M., Barham, P., Chen, J., Chen, Z., Davis, A., Dean, J., Devin, M., Ghemawat, S., Irving, G., Isard, M., et al.: Tensorflow: a system for large-scale machine learning. In: Proceedings of the 12th USENIX Symposium on Operating Systems Design and Implementation (OSDI), vol. 16, pp. 265–283 (2016)
2. AG, A.: Audi Autonomous Driving Cup (2019). https://www.audi-autonomous-driving-cup. com
3. Agarwal, S., Vora, A., Pandey, G., Williams, W., Kourous, H., McBride, J.: Ford multi-AV seasonal dataset (2020)
4. Agrawal, N., Prabhakaran, V., Wobber, T., Davis, J.D., Manasse, M., Panigrahy, R.: Design tradeoffs for SSD performance. In: Usenix Technical Conference, Boston, pp. 57–70 (2008)
5. Al-Dweik, A., Muresan, R., Mayhew, M., Lieberman, M.: Iot-based multifunctional scalable real-time enhanced road side unit for intelligent transportation systems. In: 2017 IEEE 30th Canadian Conference on Electrical and Computer Engineering (CCECE), pp. 1–6. IEEE, New York (2017)
6. Alam, M.G.R., Hassan, M.M., Uddin, M.Z., Almogren, A., Fortino, G.: Autonomic computation of floading in mobile edge for IoT applications. Fut. Gener. Comput. Syst. **90**, 149–157 (2019)
7. Ali, I., Hassan, A., Li, F.: Authentication and privacy schemes for vehicular ad hoc networks (VANETs): a survey. Veh. Commun. **16**, 45–61 (2019). https://doi.org/10.1016/j.vehcom. 2019.02.002. http://www.sciencedirect.com/science/article/pii/S221420961830319X
8. Alletto, S., Palazzi, A., Solera, F., Calderara, S., Cucchiara, R.: DR (eye)VE: a dataset for attention-based tasks with applications to autonomous and assisted driving. In: Proceedings of the IEEE Conference on Computer Vision and Pattern Recognition Workshops, pp. 54–60 (2016)
9. Allou, S., Youcef, Z., Aissa, B.: Fuzzy logic controller for autonomous vehicle path tracking. pp. 328–333 (2017). https://doi.org/10.1109/STA.2017.8314969
10. Amarasinghe, M., Kottegoda, S., Arachchi, A.L., Muramudalige, S., Bandara, H.M.N.D., Azeez, A.: Cloud-based driver monitoring and vehicle diagnostic with OBD2 telematics. In: 2015 Fifteenth International Conference on Advances in ICT for Emerging Regions (ICTer), pp. 243–249 (2015). https://doi.org/10.1109/ICTER.2015.7377695
11. An in-depth look at google's first tensor processing unit (TPU) (2019). https://cloud.google. com/blog/products/gcp/an-in-depth-look-at-googles-first-tensor-processing-unit-tpu
12. Ananthanarayanan, G., Bahl, P., Bodík, P., Chintalapudi, K., Philipose, M., Ravindranath, L., Sinha, S.: Real-time video analytics: the killer app for edge computing. Computer **50**(10), 58–67 (2017)

© The Author(s), under exclusive license to Springer Nature Switzerland AG 2021
W. Shi, L. Liu, *Computing Systems for Autonomous Driving*,
https://doi.org/10.1007/978-3-030-81564-6

13. Andrews, J.G., Buzzi, S., Choi, W., Hanly, S.V., Lozano, A., Soong, A.C., Zhang, J.C.: What will 5G be? IEEE J. Select. Areas Commun. **32**(6), 1065–1082 (2014)
14. Antinyan, V.: Revealing the complexity of automotive software. In: Proceedings of the 28th ACM Joint Meeting on European Software Engineering Conference and Symposium on the Foundations of Software Engineering, pp. 1525–1528 (2020)
15. Apollo (2017). http://apollo.auto/index.html
16. Apollo, B.: Apollo synthetic dataset. [EB/OL]. https://apollo.auto/synthetic.html. Accessed 24 Jan 2021
17. Arnautov, S., Trach, B., Gregor, F., Knauth, T., Martin, A., Priebe, C., Lind, J., Muthuku-maran, D., O'Keeffe, D., Stillwell, M.L., Goltzsche, D., Eyers, D., Kapitza, R., Pietzuch, P., Fetzer, C.: SCONE: secure linux containers with intel SGX. In: 12th USENIX Symposium on Operating Systems Design and Implementation (OSDI 16), pp. 689–703. USENIX Association, Savannah, GA (2016)
18. Association, G.A., et al.: The case for cellular V2X for safety and cooperative driving. White Paper, November 16 (2016)
19. AUTOSAR: AUTOSAR website. [Online]. https://www.autosar.org/
20. AUTOWARE (2020). https://www.autoware.org. Online. Accessed 01 Dec 2020
21. Automated driving toolbox. http://bit.ly/ToolboxMATLAB (2020). Online. Accessed 13 Dec 2020
22. Automated driving toolbox reference applications. http://bit.ly/AutomatedDrivingToolbox. Online. Accessed 30 Dec 2020
23. Automated driving toolbox ground truth labeling. http://bit.ly/GroundTruthLabeling. Online. Accessed 30 Dec 2020
24. Automotive edge computing consortium (2018). https://aecc.org/
25. Autonomous cars' big problem: The energy consumption of edge processing reduces a car's mileage with up to 30%. (2019). https://medium.com/@teraki/energy-consumption-required-by-edge-computing-reduces-a-autonomous-cars-mileage-with-up-to-30-46b6764ea1b7
26. Automated Driving Toolbox: Design, simulate, and test ADAS and autonomous driving systems (2020). https://www.mathworks.com/products/automated-driving.html
27. AVL DRIVINGCUBE: A new way to speed up the validation and approval process of ADAS/AD systems (2020). https://www.avl.com/pos-test/-/asset_publisher/gkkFgTqjTyJh/content/avl-drivingcube
28. Avola, D., Foresti, G.L., Cinque, L., Massaroni, C., Vitale, G., Lombardi, L.: A multipurpose autonomous robot for target recognition in unknown environments. In: 2016 IEEE 14th International Conference on Industrial Informatics (INDIN), pp. 766–771 (2016). https://doi.org/10.1109/INDIN.2016.7819262
29. Baidu Apollo (2020). http://bit.ly/ApolloAuto. Online. Accessed 013 Dec 2020
30. Baidu: Apollo Cyber. [Online]. https://github.com/ApolloAuto/apollo/tree/master/cyber
31. Baidu: Apollo Open Platform (2018). http://apollo.auto/index.html
32. Banzi, M., Shiloh, M.: Getting Started with Arduino: The Open Source Electronics Prototyping Platform. Maker Media, Inc., Sebastopol (2014)
33. Bao, L., Fan, L., Miao, Z.: Real-time simulation of electric vehicle battery charging systems. In: 2018 North American Power Symposium (NAPS), pp. 1–6. IEEE, New York (2018)
34. Barnes, D., Gadd, M., Murcutt, P., Newman, P., Posner, I.: The oxford radar robotcar dataset: A radar extension to the oxford robotcar dataset. In: 2020 IEEE International Conference on Robotics and Automation (ICRA), pp. 6433–6438. IEEE, New York (2020)
35. Barwacz, A.: Self-driving work zone vehicles enhance safety (2019). https://www.gpsworld.com/self-driving-work-zone-vehicles-enhance-safety/
36. Baskaran, A., Talebpour, A., Bhattacharyya, S.: End-to-End Drive By-Wire PID Lateral Control of an Autonomous Vehicle, pp. 365–376 (2020). https://doi.org/10.1007/978-3-030-32520-6_29
37. Bast, H., Delling, D., Goldberg, A., Müller-Hannemann, M., Pajor, T., Sanders, P., Wagner, D., Werneck, R.: Route planning in transportation networks (2015). arXiv:1504.05140v1 [cs.DS]

38. Behrendt, K.: Boxy vehicle detection in large images. In: Proceedings of the IEEE/CVF International Conference on Computer Vision Workshops, pp. 0–0 (2019)
39. Behrendt, K., Novak, L., Botros, R.: A deep learning approach to traffic lights: detection, tracking, and classification. In: 2017 IEEE International Conference on Robotics and Automation (ICRA), pp. 1370–1377. IEEE, New York (2017)
40. BeiDou Navigation Satellite System (2019). http://en.beidou.gov.cn/
41. Berger, C., Rumpe, B.: Autonomous driving-5 years after the urban challenge: the anticipatory vehicle as a cyber-physical system (2014). Preprint. arXiv:1409.0413
42. Bernardin, F., Bremond, R., Ledoux, V., Pinto, M., Lemonnier, S., Cavallo, V., Colomb, M.: Measuring the effect of the rainfall on the windshield in terms of visual performance. Accident Anal. Prev. **63**, 83–88 (2014)
43. Bernstein, D.J., Lange, T.: Post-quantum cryptography. Nature **549**, 188–194 (2017)
44. Besl, P., McKay, H.: A method for registration of 3-D shapes. IEEE Trans. Patt. Anal. Mach. Intell. **14**, 239–256 (1992). https://doi.org/10.1109/34.121791
45. Bewley, A., Ge, Z., Ott, L., Ramos, F., Upcroft, B.: Simple online and realtime tracking. In: 2016 IEEE International Conference on Image Processing (ICIP), pp. 3464–3468. IEEE, New York (2016)
46. Biber, P., Straßer, W.: The normal distributions transform: a new approach to laser scan matching, vol.3, pp. 2743–2748 (2003). https://doi.org/10.1109/IROS.2003.1249285
47. Bienia, C., Kumar, S., Singh, J.P., Li, K.: The PARSEC benchmark suite: characterization and architectural implications. In: Proceedings of the 17th International Conference on Parallel Architectures and Compilation Techniques, pp. 72–81. ACM, New York (2008)
48. Binas, J., Neil, D., Liu, S.C., Delbruck, T.: DDD17: end-to-end DAVIS driving dataset (2017). Preprint. arXiv:1711.01458
49. Birdsall, M.: Google and ITE: The road ahead for self-driving cars. Institute of Transportation Engineers. ITE J. **84**(5), 36 (2014)
50. Blaze, M.: A cryptographic file system for UNIX. In: Proceedings of the 1st ACM Conference on Computer and Communications Security, CCS '93, pp. 9–16. Association for Computing Machinery, New York (1993). https://doi.org/10.1145/168588.168590
51. Bochkovskiy, A., Wang, C.Y., Liao, H.Y.M.: YOLOv4: optimal speed and accuracy of object detection (2020). Preprint. arXiv:2004.10934
52. Bojarski, M., Del Testa, D., Dworakowski, D., Firner, B., Flepp, B., Goyal, P., Jackel, L.D., Monfort, M., Muller, U., Zhang, J., et al.: End to end learning for self-driving cars (2016). Preprint. arXiv:1604.07316
53. Bonnefoi, F., Bellotti, F., Scendzielorz, T., Visintainer, F.: SAFESPOT applications for infrasructurebased co-operative road safety. In: 14th World Congress and Exhibition on Intelligent Transport Systems and Services, pp. 1–8 (2007)
54. Bonomi, F., Milito, R., Zhu, J., Addepalli, S.: Fog computing and its role in the internet of things. In: Proceedings of the First Edition of the MCC Workshop on Mobile Cloud Computing, MCC '12, pp. 13–16. ACM, New York (2012). https://doi.org/10.1145/2342509.2342513
55. Borkar, A., Hayes, M., Smith, M.T.: A novel lane detection system with efficient ground truth generation. IEEE Trans. Intell. Transport. Syst. **13**(1), 365–374 (2011)
56. Bort, J.: The 'Google Brain' is a real thing but very few people have seen it. (2016). http://www.businessinsider.com/what-isgoogle-brain-2016-9
57. Bowers, K., Buscher, V., Dentten, R., Edwards, M., England, J., Enzer, M., Schooling, J., Parlikad, A.: Smart infrastructure getting more from strategic assets (2017). Corpus ID: 188185932
58. Bowne, B.F., Baker, N.R., Marzinzik, D.L., Riley, M.E., Christopulos, N.U., Fields, B.M., Wilson, J.L., Wilkerson, B.T., Thurber, D.W., et al.: Methods to determine a vehicle insurance premium based on vehicle operation data collected via a mobile device (2013). US Patent App. 13/763,231
59. Bresson, G., Alsayed, Z., Yu, L., Glaser, S.: Simultaneous localization and mapping: a survey of current trends in autonomous driving. IEEE Trans. Intell. Veh. **PP**, 1–1 (2017). https://doi.org/10.1109/TIV.2017.2749181

60. Broderick, J.A., Tilbury, D.M., Atkins, E.M.: Optimal coverage trajectories for a UGV with tradeoffs for energy and time. Autonom. Rob. **36**(3), 257–271 (2014). https://doi.org/10.1007/s10514-013-9348-x

61. Broggi, A., Buzzoni, M., Debattisti, S., Grisleri, P., Laghi, M.C., Medici, P., Versari, P.: Extensive tests of autonomous driving technologies. IEEE Trans. Intell. Transport. Syst. **14**(3), 1403–1415 (2013)

62. Brostow, G.J., Fauqueur, J., Cipolla, R.: Semantic object classes in video: A high-definition ground truth database. Patt. Recogn. Lett. **30**(2), 88–97 (2009)

63. Butler, R.W., Finelli, G.B.: The infeasibility of experimental quantification of life-critical software reliability. In: Proceedings of the Conference on Software for Critical Systems, pp. 66–76 (1991)

64. Caesar, H., Bankiti, V., Lang, A.H., Vora, S., Liong, V.E., Xu, Q., Krishnan, A., Pan, Y., Baldan, G., Beijbom, O.: nuScenes: A multimodal dataset for autonomous driving. In: Proceedings of the IEEE/CVF conference on Computer Vision and Pattern Recognition, pp. 11621–11631 (2020)

65. Cai, Z., Fan, Q., Feris, R.S., Vasconcelos, N.: A unified multi-scale deep convolutional neural network for fast object detection. In: European Conference on Computer Vision, pp. 354–370. Springer, New York (2016)

66. Camara, F., Bellotto, N., Cosar, S., Weber, F., Nathanael, D., Althoff, M., Wu, J., Ruenz, J., Dietrich, A., Markkula, G., et al.: Pedestrian models for autonomous driving part II: high-level models of human behavior. In: IEEE Transactions on Intelligent Transportation Systems (2020)

67. Cao, J., Xu, L., Abdallah, R., Shi, W.: EdgeOS_H: A home operating system for internet of everything. In: 2017 IEEE 37th International Conference on Distributed Computing Systems (ICDCS), pp. 1756–1764 (2017). https://doi.org/10.1109/ICDCS.2017.325

68. Cao, Y., Xiao, C., Cyr, B., Zhou, Y., Park, W., Rampazzi, S., Chen, Q.A., Fu, K., Mao, Z.M.: Adversarial sensor attack on LiDAR-based perception in autonomous driving. In: Proceedings of the 2019 ACM SIGSAC Conference on Computer and Communications Security, CCS '19, p. 2267–2281. Association for Computing Machinery, New York, (2019). https://doi.org/10.1145/3319535.3339815

69. Demo of prius in ros/gazebo. https://github.com/osrf/car_demo (2019). Online. Accessed 01 Dec 2020

70. CarSim ADAS: Moving objects and sensors. http://bit.ly/CarSimMO (2020). Online. Accessed 13 Dec 2020

71. Cebe, M., Erdin, E., Akkaya, K., Aksu, H., Uluagac, S.: Block4forensic: an integrated lightweight blockchain framework for forensics applications of connected vehicles. IEEE Commun. Mag. **56**(10), 50–57 (2018). https://doi.org/10.1109/MCOM.2018.1800137

72. Chakradhar, S., Sankaradas, M., Jakkula, V., Cadambi, S.: A dynamically configurable coprocessor for convolutional neural networks. In: International Symposium on Computer Architecture, pp. 247–257 (2010)

73. Chang, M.F., Lambert, J., Sangkloy, P., Singh, J., Bak, S., Hartnett, A., Wang, D., Carr, P., Lucey, S., Ramanan, D., et al.: Argoverse: 3d tracking and forecasting with rich maps. In: Proceedings of the IEEE/CVF Conference on Computer Vision and Pattern Recognition, pp. 8748–8757 (2019)

74. Che, Z., Li, G., Li, T., Jiang, B., Shi, X., Zhang, X., Lu, Y., Wu, G., Liu, Y., Ye, J.: d^2-city: A large-scale dashcam video dataset of diverse traffic scenarios (2019). Preprint. arXiv:1904.01975

75. Chen, C., Seff, A., Kornhauser, A., Xiao, J.: Deepdriving: learning affordance for direct perception in autonomous driving. In: The IEEE International Conference on Computer Vision (ICCV), pp. 2722–2730 (2015). https://doi.org/10.1109/ICCV.2015.312

76. Chen, L.C., Papandreou, G., Schroff, F., Adam, H.: Rethinking atrous convolution for semantic image segmentation (2017). Preprint. arXiv:1706.05587

77. Chen, M., Hao, Y.: Task offloading for mobile edge computing in software defined ultra-dense network. IEEE J. Select. Areas Commun. **36**(3), 587–597 (2018)

78. Chen, X., Kundu, K., Zhang, Z., Ma, H., Fidler, S., Urtasun, R.: Monocular 3D object detection for autonomous driving. In: Proceedings of the IEEE Conference on Computer Vision and Pattern Recognition, pp. 2147–2156 (2016)
79. Chen, X., Ma, H., Wan, J., Li, B., Xia, T.: Multi-view 3d object detection network for autonomous driving. In: The IEEE Conference on Computer Vision and Pattern Recognition (CVPR), pp. 6526–6534 (2017). https://doi.org/10.1109/CVPR.2017.691
80. Chen, Y., Gehlen, G., Jodlauk, G., Sommer, C., Görg, C.: A flexible application layer protocol for automotive communications in cellular networks. In: 15th World Congress on Intelligent Transportation Systems (ITS 2008), New York City, NY (2008)
81. Chen, Y., Wang, J., Li, J., Lu, C., Luo, Z., Xue, H., Wang, C.: Lidar-video driving dataset: learning driving policies effectively. In: Proceedings of the IEEE Conference on Computer Vision and Pattern Recognition, pp. 5870–5878 (2018)
82. Chen, Y.H., Emer, J., Sze, V.: Eyeriss: a spatial architecture for energy-efficient dataflow for convolutional neural networks. In: ACM SIGARCH Computer Architecture News, vol. 44, pp. 367–379. IEEE Press, New York (2016)
83. Cheng, K.W.E., Divakar, B., Wu, H., Ding, K., Ho, H.F.: Battery-management system (BMS) and SOC development for electrical vehicles. IEEE Trans. Veh. Technol. 60(1), 76–88 (2011)
84. Chevreuil, M.: IVHW: An inter-vehicle hazard warning system concept within the DEUFRAKO program. In: e-safety Congress and Exhibition, 2002, Lyon (2002)
85. Chiu, K.Y., Lin, S.F.: Lane detection using color-based segmentation. In: IEEE Proceedings. Intelligent Vehicles Symposium, 2005, pp. 706–711. IEEE, New York (2005)
86. Choi, W., Nam, H.S., Kim, B., Ahn, C.: Model Predictive Control for Evasive Steering of Autonomous Vehicle, pp. 1252–1258 (2020). https://doi.org/10.1007/978-3-030-38077-9_144
87. Chollet, F.: Xception: deep learning with depthwise separable convolutions, pp. 1610–2357 (2017). arXiv. Preprint.
88. Claudio, S., Giancarlo, F.: A simulation-driven methodology for IoT data mining based on edge computing, in *ACM Transactions on Internet Technology (TOIT)* (2020)
89. Clemons, J., Zhu, H., Savarese, S., Austin, T.: MEVBench: A mobile computer vision benchmarking suite. In: 2011 IEEE International Symposium on Workload Characterization (IISWC), pp. 91–102. IEEE, New York (2011)
90. Coates, A., Huval, B., Wang, T., Wu, D., Catanzaro, B., Andrew, N.: Deep learning with cots HPC systems. In: International Conference on Machine Learning, pp. 1337–1345 (2013)
91. Collaborative learning on the edges: A case study on connected vehicles. In: 2nd USENIX Workshop on Hot Topics in Edge Computing (HotEdge 19). USENIX Association, Renton, WA (2019). https://www.usenix.org/conference/hotedge19/presentation/lu
92. of Concerned Scientists, U.: Electric vehicle batteries: materials, cost, lifespan (2018). https://www.ucsusa.org/resources/ev-batteries
93. Continental ARS4-A 77GHz Radar (2017). https://www.systemplus.fr/reverse-costing-reports/continental-ars4-a-77ghz-radar/
94. Cordts, M., Omran, M., Ramos, S., Rehfeld, T., Enzweiler, M., Benenson, R., Franke, U., Roth, S., Schiele, B.: The cityscapes dataset for semantic urban scene understanding. In: Proceedings of the IEEE Conference on Computer Vision and Pattern Recognition, pp. 3213–3223 (2016)
95. Courbariaux, M., Bengio, Y., David, J.P.B.: Training deep neural networks with binary weights during propagations (2015). arxiv. Preprint. arXiv:1511.00363
96. Dalal, N., Triggs, B.: Histograms of oriented gradients for human detection. In: 2005 IEEE Computer Society Conference on Computer Vision and Pattern Recognition (CVPR'05), vol. 1, pp. 886–893. IEEE, New York (2005)
97. Danescu, R., Nedevschi, S.: Probabilistic lane tracking in difficult road scenarios using stereovision. IEEE Trans. Intell. Transport. Syst. 10(2), 272–282 (2009)
98. Data storage is the key to autonomous vehicles' future (2019). https://iotnowtransport.com/2019/02/12/71015-data-storage-key-autonomous-vehicles-future/. Accessed 30 Dec 2019

99. De La Fortelle, A., Qian, X., Diemer, S., Grégoire, J., Moutarde, F., Bonnabel, S., Marjovi, A., Martinoli, A., Llatser, I., Festag, A., et al.: Network of automated vehicles: the AutoNet 2030 vision. In: ITS World Congress (2014)

100. Dean, J., Ghemawat, S.: Mapreduce: simplified data processing on large clusters. Commun. ACM **51**(1), 107–113 (2008)

101. Deng, J., Dong, W., Socher, R., Li, L.J., Li, K., Fei-Fei, L.: ImageNet: a large-scale hierarchical image database. In: 2009 IEEE Conference on Computer Vision and Pattern Recognition, pp. 248–255. IEEE, New York (2009)

102. Deng, R., Di, B., Song, L.: Cooperative collision avoidance for overtaking maneuvers in cellular V2X-based autonomous driving. IEEE Trans. Veh. Technol. **68**(5), 4434–4446 (2019)

103. Denil, M., Shakibi, B., Dinh, L., De Freitas, N., et al.: Predicting parameters in deep learning. In: Advances in Neural Information Processing Systems, pp. 2148–2156 (2013)

104. Denton, E.L., Zaremba, W., Bruna, J., LeCun, Y., Fergus, R.: Exploiting linear structure within convolutional networks for efficient evaluation. In: Advances in Neural Information Processing Systems, pp. 1269–1277 (2014)

105. Deschaud, J.E.: IMLS-SLAM: scan-to-model matching based on 3D data. pp. 2480–2485 (2018). https://doi.org/10.1109/ICRA.2018.8460653

106. DEUFRAKO (2010). http://deufrako.org/web/index.php. http://deufrako.org/web/index.php

107. Deusch, H., Wiest, J., Reuter, S., Szczot, M., Konrad, M., Dietmayer, K.: A random finite set approach to multiple lane detection. In: 2012 15th International IEEE Conference on Intelligent Transportation Systems, pp. 270–275. IEEE, New York (2012)

108. Di, S., Cappello, F.: Fast error-bounded lossy HPC data compression with SZ. In: 2016 IEEE International Parallel and Distributed Processing Symposium (IPDPS), pp. 730–739 (2016). https://doi.org/10.1109/IPDPS.2016.11

109. Dias, J.A., Rodrigues, J.J., Kumar, N., Saleem, K.: Cooperation strategies for vehicular delay-tolerant networks. IEEE Commun. Mag. **53**(12), 88–94 (2015)

110. Diffenderfer, J., Fox, A., Hittinger, J., Sanders, G., Lindstrom, P.: Error analysis of ZFP compression for floating-point data. SIAM J. Sci. Comput. **41**, A1867–A1898 (2019). https://doi.org/10.1137/18M1168832

111. Dimitrakopoulos, G., Demestichas, P.: Intelligent transportation systems. IEEE Veh. Technol. Mag. **5**(1), 77–84 (2010)

112. Ding, W., Yan, Z., Deng, R.: Privacy-preserving data processing with flexible access control. IEEE Trans. Depend. Secure Comput. 1–1 (2017). https://doi.org/10.1109/TDSC.2017.2786247

113. Dixit, S., Montanaro, U., Fallah, S., Dianati, M., Oxtoby, D., Mizutani, T., Mouzakitis, A.: Trajectory planning for autonomous high-speed overtaking using MPC with terminal set constraints (2018). https://doi.org/10.1109/ITSC.2018.8569529

114. DJI: RoboMaster Robotics Competition (2019). https://www.robomaster.com

115. Dogru, S., Marques, L.: Energy efficient coverage path planning for autonomous mobile robots on 3D terrain. In: 2015 IEEE International Conference on Autonomous Robot Systems and Competitions, pp. 118–123 (2015). https://doi.org/10.1109/ICARSC.2015.23

116. Dollár, P., Wojek, C., Schiele, B., Perona, P.: Pedestrian detection: a benchmark. In: 2009 IEEE Conference on Computer Vision and Pattern Recognition, pp. 304–311. IEEE, New York (2009)

117. Dong, Z., Shi, W., Tong, G., Yang, K.: Collaborative autonomous driving: vision and challenges (2020). https://doi.org/10.1109/MetroCAD48866.2020.00010

118. dos Santos Lima, F.D., Amaral, G.M.R., de Moura Leite, L.G., Gomes, J.P.P., de Castro Machado, J.: Predicting failures in hard drives with LSTM networks. In: Proceedings of the 2017 Brazilian Conference on Intelligent Systems (BRACIS), pp. 222–227. IEEE, New York (2017)

119. Dosovitskiy, A., Ros, G., Codevilla, F., Lopez, A., Koltun, V.: CARLA: an open urban driving simulator (2017). Preprint. arXiv:1711.03938

120. DOT, U.: Intelligent transportation systems-joint program (2019)

121. Du, D., Qi, Y., Yu, H., Yang, Y., Duan, K., Li, G., Zhang, W., Huang, Q., Tian, Q.: The unmanned aerial vehicle benchmark: object detection and tracking. In: Proceedings of the European Conference on Computer Vision (ECCV), pp. 370–386 (2018)

122. Dua, R., Raja, A.R., Kakadia, D.: Virtualization vs containerization to support paas. In: 2014 IEEE International Conference on Cloud Engineering, pp. 610–614 (2014). https://doi.org/10.1109/IC2E.2014.41

123. Durmuş, H., Güneş, E.O., Kırcı, M., Üstündağ, B.B.: The design of general purpose autonomous agricultural mobile-robot: "AGROBOT". In: 2015 Fourth International Conference on Agro-Geoinformatics (Agro-geoinformatics), pp. 49–53 (2015). https://doi.org/10.1109/Agro-Geoinformatics.2015.7248088

124. Durrant-Whyte, H., Bailey, T.: Simultaneous localization and mapping: part I. IEEE Robot. Autom. Mag. **13**(2), 99–110 (2006)

125. Emmanuel, I.: Fuzzy logic-based control for autonomous vehicle: a survey. Int. J. Educ. Manage. Eng. **7**, 41–49 (2017). https://doi.org/10.5815/ijeme.2017.02.05

126. Empowering Safe Autonomous Driving (2020). https://www.dspace.com/en/inc/home/applicationfields/our_solutions_for/driver_assistance_systems.cfm

127. Enabling Next Generation ADAS and AD Systems (2020). https://www.xilinx.com/products/silicon-devices/soc/xa-zynq-ultrascale-mpsoc.html

128. Endres, F., Hess, J., Sturm, J., Cremers, D., Burgard, W.: 3-D mapping with an RGB-D camera. IEEE Trans. Robot. **30**, 177–187 (2014). https://doi.org/10.1109/TRO.2013.2279412

129. Engel, J., Schoeps, T., Cremers, D.: LSD-SLAM: large-scale direct monocular SLAM, pp. 1–16 (2014). https://doi.org/10.1007/978-3-319-10605-2_54

130. Ensuring American Leadership in Automated Vehicle Technologies: Automated Vehicles 4.0 (2020). https://www.transportation.gov/av/4

131. Ernst, J.M., Michaels, A.J.: Lin bus security analysis. In: IECON 2018 - 44th Annual Conference of the IEEE Industrial Electronics Society, pp. 2085–2090 (2018)

132. Eshraghi, N., Liang, B.: Joint offloading decision and resource allocation with uncertain task computing requirement. In: IEEE INFOCOM 2019-IEEE Conference on Computer Communications, pp. 1414–1422. IEEE, New York (2019)

133. Everingham, M., Van Gool, L., Williams, C.K., Winn, J., Zisserman, A.: The pascal visual object classes (voc) challenge. Int. J. Comput. Vis. **88**(2), 303–338 (2010)

134. Fadaie, J.: The state of modeling, simulation, and data utilization within industry: an autonomous vehicles perspective (2019). Preprint. arXiv:1910.06075

135. Fairfield, N., Urmson, C.: Traffic light mapping and detection. In: 2011 IEEE International Conference on Robotics and Automation, pp. 5421–5426. IEEE,, New York (2011)

136. Fan, R., Jiao, J., Ye, H., Yu, Y., Pitas, I., Liu, M.: Key ingredients of self-driving cars (2019). Preprint. arXiv:1906.02939

137. Felzenszwalb, P., McAllester, D., Ramanan, D.: A discriminatively trained, multiscale, deformable part model. In: 2008 IEEE Conference on Computer Vision and Pattern Recognition, pp. 1–8. IEEE, New York (2008)

138. Feng, Z., George, S., Harkes, J., Pillai, P., Klatzky, R., Satyanarayanan, M.: Edge-based discovery of training data for machine learning. In: 2018 IEEE/ACM Symposium on Edge Computing (SEC), pp. 145–158. IEEE, New York (2018)

139. Festag, A., Fußler, H., Hartenstein, H., Sarma, A., Schmitz, R.: FleetNet: bringing car-to-car communication into the real world. Computer **4**(L15), 16 (2004)

140. Festag, A., Noecker, G., Strassberger, M., Lübke, A., Bochow, B., Torrent-Moreno, M., Schnaufer, S., Eigner, R., Catrinescu, C., Kunisch, J.: 'NoW–Network on Wheels': project objectives, technology and achievements (2008)

141. Figueiredo, M.C., Rossetti, R.J., Braga, R.A., Reis, L.P.: An approach to simulate autonomous vehicles in urban traffic scenarios. In: 2009 12th International IEEE Conference on Intelligent Transportation Systems, pp. 1–6. IEEE, New York (2009)

142. Filgueira, A., González-Jorge, H., Lagüela, S., Díaz-Vilariño, L., Arias, P.: Quantifying the influence of rain in LiDAR performance. Measurement **95**, 143–148 (2017)

143. Flood of data will get generated in autonomous cars (2020). https://autotechreview.com/features/flood-of-data-will-get-generated-in-autonomous-cars. Accessed 18 Feb 2020

144. Fogue, M., Sanguesa, J.A., Martinez, F.J., Marquez-Barja, J.M.: Improving roadside unit deployment in vehicular networks by exploiting genetic algorithms. Appl. Sci. **8**(1), 86 (2018)

145. Ford: SYNC (2018). https://www.ford.com/technology/sync/

146. Ford will have a fully autonomous vehicle in operation by 2021 (2018). https://corporate.ford.com/innovation/autonomous-2021.html

147. Forecast, G.: Cisco visual networking index: Global mobile data traffic forecast update 2017–2022. Update **2017**, 2022 (2019)

148. Fotouhi, A., Auger, D.J., Propp, K., Longo, S., Wild, M.: A review on electric vehicle battery modelling: From lithium-ion toward lithium–sulphur. Renew. Sustain. Energy Rev. **56**, 1008–1021 (2016)

149. Franz, W., Hartenstein, H., Mauve, M.: Inter-vehicle-communications based on ad hoc networking principles: the FleetNet project. Universitätsverlag Karlsruhe Karlsruhe (2005)

150. Fregin, A., Muller, J., Krebel, U., Dietmayer, K.: The DriveU traffic light dataset: introduction and comparison with existing datasets. In: 2018 IEEE International Conference on Robotics and Automation (ICRA), pp. 3376–3383. IEEE, New York (2018)

151. Friedman, J.H.: Stochastic gradient boosting. Comput. Stat. Data Anal. **38**(4), 367–378 (2002)

152. Fu, C.Y., Liu, W., Ranga, A., Tyagi, A., Berg, A.C.: DSSD: deconvolutional single shot detector (2017). Preprint. arXiv:1701.06659

153. Funke, J., Brown, M., Erlien, S.M., Gerdes, J.C.: Collision avoidance and stabilization for autonomous vehicles in emergency scenarios. IEEE Trans. Control Syst. Technol. **25**(4), 1204–1216 (2017)

154. Furlong, M., Quinn, A., Flinn, J.: The case for determinism on the edge. In: 2nd {USENIX} Workshop on Hot Topics in Edge Computing (HotEdge 19) (2019)

155. Gaidon, A., Wang, Q., Cabon, Y., Vig, E.: Virtual worlds as proxy for multi-object tracking analysis. In: 2016 IEEE Conference on Computer Vision and Pattern Recognition (CVPR) (2016)

156. García, G.J., Jara, C.A., Pomares, J., Alabdo, A., Poggi, L.M., Torres, F.: A survey on fpga-based sensor systems: towards intelligent and reconfigurable low-power sensors for computer vision, control and signal processing. Sensors **14**(4), 6247–6278 (2014)

157. Garcia-Garcia, A., Orts-Escolano, S., Oprea, S., Villena-Martinez, V., Garcia-Rodriguez, J.: A review on deep learning techniques applied to semantic segmentation (2017). Preprint. arXiv:1704.06857

158. Garg, S., Singh, A., Kaur, K., Aujla, G.S., Batra, S., Kumar, N., Obaidat, M.S.: Edge computing-based security framework for big data analytics in VANETs. IEEE Network **33**(2), 72–81 (2019)

159. Garnett, N., Cohen, R., Pe'er, T., Lahav, R., Levi, D.: 3D-LaneNet: end-to-end 3D multiple lane detection. In: Proceedings of the IEEE International Conference on Computer Vision, pp. 2921–2930 (2019)

160. Gawel, A., Don, C., Siegwart, R., Nieto, J., Cadena, C.: X-View: graph-based semantic multi-view localization. IEEE Robot. Autom. Lett. **3**, 1687–1694 (2018). https://doi.org/10.1109/LRA.2018.2801879

161. Geiger, A., Lauer, M., Wojek, C., Stiller, C., Urtasun, R.: 3d traffic scene understanding from movable platforms. IEEE Trans. Patt. Anal. Mach. Intell. **36**(5), 1012–1025 (2014)

162. Geiger, A., Lenz, P., Stiller, C., Urtasun, R.: Vision meets robotics: the KITTI dataset. Int. J. Robot. Res. **32**(11), 1231–1237 (2013)

163. Geiger, A., Lenz, P., Urtasun, R.: Are we ready for autonomous driving? The KITTI vision benchmark suite. In: 2012 IEEE Conference on Computer Vision and Pattern Recognition, pp. 3354–3361. IEEE, New York (2012)

164. Geng, X., Liang, H., Yu, B., Zhao, P., He, L., Huang, R.: A scenario-adaptive driving behavior prediction approach to urban autonomous driving. Appl. Sci. **7**(4), 426 (2017)

165. Gentner, C., Jost, T., Wang, W., Zhang, S., Dammann, A., Fiebig, U.C.: Multipath assisted positioning with simultaneous localization and mapping. IEEE Trans. Wirel. Commun. **15**(9), 6104–6117 (2016)

166. Gerla, M., Lee, E.K., Pau, G., Lee, U.: Internet of vehicles: from intelligent grid to autonomous cars and vehicular clouds. In: 2014 IEEE World Forum on Internet of Things (WF-IoT), pp. 241–246. IEEE, New York (2014)

167. Geyer, J., Kassahun, Y., Mahmudi, M., Ricou, X., Durgesh, R., Chung, A.S., Hauswald, L., Pham, V.H., Mühlegg, M., Dorn, S., Fernandez, T., Jänicke, M., Mirashi, S., Savani, C., Sturm, M., Vorobiov, O., Oelker, M., Garreis, S., Schuberth, P.: A2D2: Audi Autonomous Driving Dataset (2020). https://www.a2d2.audi

168. Ghafoorian, M., Nugteren, C., Baka, N., Booij, O., Hofmann, M.: EL-GAN: embedding loss driven generative adversarial networks for lane detection. In: Proceedings of the European Conference on Computer Vision (ECCV) (2018)

169. Ghosh, A., Paranthaman, V.V., Mapp, G., Gemikonakli, O., Loo, J.: Enabling seamless V2I communications: toward developing cooperative automotive applications in VANET systems. IEEE Commun. Mag. **53**(12), 80–86 (2015)

170. Giancola, S., Zarzar, J., Ghanem, B.: Leveraging shape completion for 3d siamese tracking. In: Proceedings of the IEEE Conference on Computer Vision and Pattern Recognition, pp. 1359–1368 (2019)

171. Girshick, R.: Fast R-CNN. In: Proceedings of the IEEE International Conference on Computer Vision, pp. 1440–1448 (2015)

172. Girshick, R., Donahue, J., Darrell, T., Malik, J.: Rich feature hierarchies for accurate object detection and semantic segmentation. In: Proceedings of the IEEE Conference on Computer Vision and Pattern Recognition, pp. 580–587 (2014)

173. Girshick, R., Donahue, J., Darrell, T., Malik, J.: Region-based convolutional networks for accurate object detection and segmentation. IEEE Trans. Patt. Anal. Mach. Intell. **38**(1), 142–158 (2015)

174. Global Autonomous Driving Market Outlook, 2018 (2018). https://www.prnewswire.com/news-releases/global-autonomous-driving-market-outlook-2018-300624588.html

175. Gomez-Miguelez, I., Garcia-Saavedra, A., Sutton, P.D., Serrano, P., Cano, C., Leith, D.J.: srsLTE: an open-source platform for LTE evolution and experimentation. In: Proceedings of the Tenth ACM International Workshop on Wireless Network Testbeds, Experimental Evaluation, and Characterization, pp. 25–32 (2016)

176. Gonzalez Bautista, D., Pérez, J., Milanes, V., Nashashibi, F.: A review of motion planning techniques for automated vehicles, In: IEEE Transactions on Intelligent Transportation Systems, pp. 1–11 (2015). https://doi.org/10.1109/TITS.2015.2498841

177. Google's Waymo invests in LiDAR technology, cuts costs by 90 percent (2017). https://arstechnica.com/cars/2017/01/googles-waymo-invests-in-lidar-technology-cuts-costs-by-90-percent/

178. Gopalan, R., Hong, T., Shneier, M., Chellappa, R.: A learning approach towards detection and tracking of lane markings. IEEE Trans. Intell. Transport. Syst. **13**(3), 1088–1098 (2012)

179. Goswami, D., Schneider, R., Masrur, A., Lukasiewycz, M., Chakraborty, S., Voit, H., Annaswamy, A.: Challenges in automotive cyber-physical systems design. In: 2012 International Conference on Embedded Computer Systems (SAMOS), pp. 346–354. IEEE, New York (2012)

180. GPS: The Global Positioning System. (2019). https://www.gps.gov/

181. GPS Accuracy (2020). https://www.gps.gov/systems/gps/performance/accuracy/

182. Graves, A., Schmidhuber, J.: Framewise phoneme classification with bidirectional LSTM and other neural network architectures. Neural Netw. **18**(5–6), 602–610 (2005)

183. Gregg, B.: Linux performance analysis and tools. Technical report, Joyent (2013)

184. Grigorescu, S., Trasnea, B., Cocias, T., Macesanu, G.: A survey of deep learning techniques for autonomous driving. J. Field Robot. **37**(3), 362–386 (2020)

185. Grisetti, G., Kümmerle, R., Stachniss, C., Burgard, W.: A tutorial on graph-based slam. IEEE Intell. Transport. Syst. Mag. **2**(4), 31–43 (2010)

186. Group, L.: LEGO Mindstorms Robot (2019). https://www.lego.com/en-us/mindstorms/build-a-robot

187. Grulich, P.M., Nawab, F.: Collaborative edge and cloud neural networks for real-time video processing. Proceedings of the VLDB Endowment **11**(12), 2046–2049 (2018)

188. Guilford, G.: Ford can only afford to give up on cars because of american protectionism (2018). https://qz.com/1262815/fords-move-to-stop-making-cars-was-enabled-by-american-protectionism/

189. Gulli, A., Pal, S.: Deep Learning with Keras. Packt Publishing Ltd, Birmingham (2017)

190. Guo, J., Hu, P., Wang, R.: Nonlinear coordinated steering and braking control of vision-based autonomous vehicles in emergency obstacle avoidance. IEEE Trans. Intell. Transport. Syst. **17**(11), 3230–3240 (2016)

191. Guo, J., Kurup, U., Shah, M.: Is it safe to drive? An overview of factors, metrics, and datasets for driveability assessment in autonomous driving. In: IEEE Transactions on Intelligent Transportation Systems (2019)

192. Han, S., Liu, X., Mao, H., Pu, J., Pedram, A., Horowitz, M.A., Dally, W.J.: EIE: efficient inference engine on compressed deep neural network. In: 2016 ACM/IEEE 43rd Annual International Symposium on Computer Architecture (ISCA), pp. 243–254. IEEE New York (2016)

193. Han, S., Mao, H., Dally, W.J.: Deep compression: compressing deep neural networks with pruning, trained quantization and Huffman coding (2015). Preprint. arXiv:1510.00149

194. Han, S., Pool, J., Tran, J., Dally, W.: Learning both weights and connections for efficient neural network. In: Advances in Neural Information Processing Systems, pp. 1135–1143 (2015)

195. Hank, P., Müller, S., Vermesan, O., Van Den Keybus, J.: Automotive ethernet: in-vehicle networking and smart mobility. In: 2013 Design, Automation & Test in Europe Conference & Exhibition (DATE), pp. 1735–1739. IEEE, New York (2013)

196. Hannun, A., Case, C., Casper, J., Catanzaro, B., Diamos, G., Elsen, E., Prenger, R., Satheesh, S., Sengupta, S., Coates, A., et al.: Deep Speech: scaling up end-to-end speech recognition (2014). Preprint. arXiv:1412.5567

197. Hart, P., Nilsson, N., Raphael, B.: A formal basis for the heuristic determination of minimum cost paths. Intell./sigart Bull. - SIGART **37**, 28–29 (1972). https://doi.org/10.1145/1056777.1056779

198. Hartenstein, H., Bochow, B., Ebner, A., Lott, M., Radimirsch, M., Vollmer, D.: Position-aware ad hoc wireless networks for inter-vehicle communications: the FleetNet project. In: Proceedings of the 2nd ACM International Symposium on Mobile Ad Hoc Networking & Computing, pp. 259–262. ACM, New York (2001)

199. Hata, A., Wolf, D.: Road marking detection using LiDAR reflective intensity data and its application to vehicle localization, pp. 584–589 (2014). https://doi.org/10.1109/ITSC.2014.6957753

200. He, K., Gkioxari, G., Dollár, P., Girshick, R.: Mask R-CNN. In: Proceedings of the IEEE International Conference on Computer Vision, pp. 2961–2969 (2017)

201. He, L., Kim, E., Shin, K.G., Meng, G., He, T.: Battery state-of-health estimation for mobile devices. In: 2017 ACM/IEEE 8th International Conference on Cyber-Physical Systems (ICCPS), pp. 51–60. IEEE, New York (2017)

202. Held, D., Thrun, S., Savarese, S.: Learning to track at 100 fps with deep regression networks. In: European Conference on Computer Vision, pp. 749–765. Springer, New York (2016)

203. Henkel, C., Bubeck, A., Xu, W.: Energy efficient dynamic window approach for local path planning in mobile service robotics. This work was conducted at the University of Auckland, Auckland, New Zealand. IFAC-PapersOnLine **49**(15), 32–37 (2016). https://doi.org/10.1016/j.ifacol.2016.07.610. http://www.sciencedirect.com/science/article/pii/S2405896316308813. 9th IFAC Symposium on Intelligent Autonomous Vehicles IAV 2016

204. HERE HD live map (2020). http://bit.ly/HERE_HDMaps. Online. Accessed 30 Dec 2020

205. Hess, W., Kohler, D., Rapp, H., Andor, D.: Real-time loop closure in 2D LiDAR SLAM, pp. 1271–1278 (2016). https://doi.org/10.1109/ICRA.2016.7487258

206. Hildebrand, D.: An architectural overview of QNX. In: USENIX Workshop on Microkernels and Other Kernel Architectures, pp. 113–126 (1992)

207. Hilgert, J., Hirsch, K., Bertram, T., Hiller, M.: Emergency path planning for autonomous vehicles using elastic band theory. In: Proceedings 2003 IEEE/ASME International Conference on Advanced Intelligent Mechatronics (AIM 2003), vol. 2, pp. 1390–1395 (2003)

208. Hillel, A.B., Lerner, R., Levi, D., Raz, G.: Recent progress in road and lane detection: a survey. Mach. Vis. Appl. **25**(3), 727–745 (2014)

209. Hiller, A., Hinsberger, A., Strassberger, M., Verburg, D.: Results from the WILLWARN project. In: 6th European Congress and Exhibition on Intelligent Transportation Systems and Services (2007)

210. Hirst, J.M., Miller, J.R., Kaplan, B.A., Reed, D.D.: Watts up? Pro ac power meter for automated energy recording. Behav. Anal. Pract. **6**(1), 82–95 (2013)

211. Hirz, M., Walzel, B.: Sensor and object recognition technologies for self-driving cars. Computer-Aided Des. Appl. **15**(4), 501–508 (2018)

212. Hnewa, M., Radha, H.: Object detection under rainy conditions for autonomous vehicles (2020). Preprint. arXiv:2006.16471

213. Hobert, L., Festag, A., Llatser, I., Altomare, L., Visintainer, F., Kovacs, A.: Enhancements of V2X communication in support of cooperative autonomous driving. IEEE Commun. Mag. **53**(12), 64–70 (2015)

214. Hochreiter, S., Schmidhuber, J.: Long short-term memory. Neural Comput. **9**(8), 1735–1780 (1997)

215. Hoffmann, G., Tomlin, C., Montemerlo, M., Thrun, S.: Autonomous automobile trajectory tracking for off-road driving: Controller design, experimental validation and racing. pp. 2296–2301 (2007). https://doi.org/10.1109/ACC.2007.4282788

216. Hommes, Q.V.E.: Review and assessment of the iso 26262 draft road vehicle-functional safety. Tech. rep., SAE Technical Paper (2012)

217. Hong Yoon, J., Lee, C.R., Yang, M.H., Yoon, K.J.: Online multi-object tracking via structural constraint event aggregation. In: Proceedings of the IEEE Conference on Computer Vision and Pattern Recognition, pp. 1392–1400 (2016)

218. Hou, Y.: Agnostic lane detection (2019). Preprint. arXiv:1905.03704

219. Hou, Y., Ma, Z., Liu, C., Loy, C.C.: Learning lightweight lane detection CNNs by self attention distillation. In: Proceedings of the IEEE International Conference on Computer Vision, pp. 1013–1021 (2019)

220. Houben, S., Stallkamp, J., Salmen, J., Schlipsing, M., Igel, C.: Detection of traffic signs in real-world images: The german traffic sign detection benchmark. In: The 2013 International Joint Conference on Neural Networks (IJCNN), pp. 1–8. IEEE, New York (2013)

221. Houston, J., Zuidhof, G., Bergamini, L., Ye, Y., Jain, A., Omari, S., Iglovikov, V., Ondruska, P.: One thousand and one hours: Self-driving motion prediction dataset (2020). arXiv:2006.14480v1 [cs.CV]

222. Howard, A.G., Zhu, M., Chen, B., Kalenichenko, D., Wang, W., Weyand, T., Andreetto, M., Adam, H.: Mobilenets: Efficient convolutional neural networks for mobile vision applications (2017). Preprint. arXiv:1704.04861

223. Hsu, Y.C., Xu, Z., Kira, Z., Huang, J.: Learning to cluster for proposal-free instance segmentation. In: 2018 International Joint Conference on Neural Networks (IJCNN), pp. 1–8. IEEE, New York (2018)

224. Huang, W., Wang, K., Lv, Y., Zhu, F.: Autonomous vehicles testing methods review. In: 2016 IEEE 19th International Conference on Intelligent Transportation Systems (ITSC), pp. 163–168 (2016). https://doi.org/10.1109/ITSC.2016.7795548

225. Huang, X., Cheng, X., Geng, Q., Cao, B., Zhou, D., Wang, P., Lin, Y., Yang, R.: The apolloscape dataset for autonomous driving. In: Proceedings of the IEEE Conference on Computer Vision and Pattern Recognition Workshops, pp. 954–960 (2018)

226. Hubmann, C., Becker, M., Althoff, D., Lenz, D., Stiller, C.: Decision making for autonomous driving considering interaction and uncertain prediction of surrounding vehicles. In: 2017 IEEE Intelligent Vehicles Symposium (IV), pp. 1671–1678. IEEE, New York (2017)

227. Hung, C.C., Ananthanarayanan, G., Bodik, P., Golubchik, L., Yu, M., Bahl, P., Philipose, M.: VideoEdge: processing camera streams using hierarchical clusters. In: 2018 IEEE/ACM Symposium on Edge Computing (SEC), pp. 115–131. IEEE, New York (2018)

228. Hur, J., Kang, S.N., Seo, S.W.: Multi-lane detection in urban driving environments using conditional random fields. In: 2013 IEEE Intelligent Vehicles Symposium (IV), pp. 1297–1302. IEEE, New York (2013)

229. Hwang, Y.K., Ahuja, N.: Gross motion planning—a survey. ACM Comput. Surv. **24**(3), 219–291 (1992)

230. Iandola, F.N., Han, S., Moskewicz, M.W., Ashraf, K., Dally, W.J., Keutzer, K.: Squeezenet: Alexnet-level accuracy with 50x fewer parameters and <0.5 mb model size (2016). Preprint. arXiv:1602.07360

231. Ilic, S., Katupitiya, J., Tordon, M.: In-vehicle data logging system for fatigue analysis of drive shaft. In: International Workshop on Robot Sensing, 2004, ROSE 2004, pp. 30–34 (2004)

232. Xilinx Inc.: Xilinx Zynq Ultrascale+ MPSoC ZCU106 Evaluation Kit (2018). https://www.xilinx.com/products/boards-and-kits/zcu106.html

233. international, S.: Taxonomy and definitions for terms related to driving automation systems for on-road motor vehicles (2016). http://standards.sae.org/j3016_201609/

234. Isento, J.N., Rodrigues, J.J., Dias, J.A., Paula, M.C., Vinel, A.: Vehicular delay-tolerant networks? A novel solution for vehicular communications. IEEE Intell. Transport. Syst. Mag. **5**(4), 10–19 (2013)

235. Jafarzadeh, H., Fleming, C.: Learning model predictive control for connected autonomous vehicles (2019). https://doi.org/10.1109/CDC40024.2019.9029830

236. Jain, A., Koppula, H.S., Raghavan, B., Soh, S., Saxena, A.: Car that knows before you do: anticipating maneuvers via learning temporal driving models. In: Proceedings of the IEEE International Conference on Computer Vision, pp. 3182–3190 (2015)

237. Jang, S.Y., Lee, Y., Shin, B., Lee, D.: Application-aware IoT camera virtualization for video analytics edge computing. In: 2018 IEEE/ACM Symposium on Edge Computing (SEC), pp. 132–144. IEEE, New York (2018)

238. Jiang, K., Yang, D., Liu, C., Zhang, T., Xiao, Z.: A flexible multi-layer map model designed for lane-level route planning in autonomous vehicles. Engineering **5**(2), 305–318 (2019). https://doi.org/10.1016/j.eng.2018.11.032. http://www.sciencedirect.com/science/article/pii/S2095809918300328

239. Jiang, L., Zhao, H., Shi, S., Liu, S., Fu, C.W., Jia, J.: Pointgroup: dual-set point grouping for 3d instance segmentation. In: Proceedings of the IEEE/CVF Conference on Computer Vision and Pattern Recognition, pp. 4867–4876 (2020)

240. Jouppi, N.: Google supercharges machine learning tasks with TPU custom chip. (2016). https://cloud.google.com/blog/products/gcp/google-supercharges-machine-learning-tasks-with-custom-chip

241. Jung, H., Min, J., Kim, J.: An efficient lane detection algorithm for lane departure detection. In: 2013 IEEE Intelligent Vehicles Symposium (IV), pp. 976–981. IEEE, New York (2013)

242. Kamkar, S.: Drive it like you hacked it: new attacks and tools to wirelessly steal cars. In: Presentation at DEFCON (2015)

243. Kang, L., Zhao, W., Qi, B., Banerjee, S.: Augmenting self-driving with remote control: Challenges and directions. In: Proceedings of the 19th International Workshop on Mobile Computing Systems & Applications, pp. 19–24. ACM, New York (2018)

244. Kang, Y., Hauswald, J., Gao, C., Rovinski, A., Mudge, T., Mars, J., Tang, L.: Neurosurgeon: collaborative intelligence between the cloud and mobile edge. ACM Sigplan Not. **52**(4), 615–629 (2017)

245. Kang, Y., Yin, H., Berger, C.: Test your self-driving algorithm: an overview of publicly available driving datasets and virtual testing environments. IEEE Trans. Intell. Veh. **4**(2), 171–185 (2019)

246. Kaplan, S., Guvensan, M.A., Yavuz, A.G., Karalurt, Y.: Driver behavior analysis for safe driving: a survey. IEEE Trans. Intell. Transport. Syst. **16**(6), 3017–3032 (2015)

247. Kar, G., Jain, S., Gruteser, M., Bai, F., Govindan, R.: Real-time traffic estimation at vehicular edge nodes. In: Proceedings of the Second ACM/IEEE Symposium on Edge Computing, p. 3. ACM, New York (2017)

248. Kar, G., Jain, S., Gruteser, M., Chen, J., Bai, F., Govindan, R.: Predriveid: pre-trip driver identification from in-vehicle data. In: Proceedings of the Second ACM/IEEE Symposium on Edge Computing, San Jose/Silicon Valley, SEC 2017, CA, October 12–14, pp. 2:1–2:12 (2017). https://doi.org/10.1145/3132211.3134462.

249. Kar, G., Jain, S., Gruteser, M., Chen, J., Bai, F., Govindan, R.: PredriveID: pre-trip driver identification from in-vehicle data. In: Proceedings of the Second ACM/IEEE Symposium on Edge Computing, pp. 2:1–2:12. ACM, New York (2017)

250. Karaman, S., Anders, A., Boulet, M., Connor, J., Gregson, K., Guerra, W., Guldner, O., Mohamoud, M., Plancher, B., Shin, R., et al.: Project-based, collaborative, algorithmic robotics for high school students: programming self-driving race cars at MIT. In: 2017 IEEE Integrated STEM Education Conference (ISEC), pp. 195–203. IEEE, New York (2017)

251. Karaman, S., Frazzoli, E.: Sampling-based algorithms for optimal motion planning. Int. J. Robot. Res. **30**, 846–894 (2011). https://doi.org/10.1177/0278364911406761

252. Karaman, S., Walter, M., Perez, A., Frazzoli, E., Teller, S.: Anytime motion planning using the RRT*. pp. 1478–1483 (2011). https://doi.org/10.1109/ICRA.2011.5980479

253. Karri, C., Jena, U.: Fast vector quantization using a bat algorithm for image compression. Int. J. Eng. Sci. Technol. **19**(2), 769–781 (2016). https://doi.org/10.1016/j.jestch.2015.11.003. http://www.sciencedirect.com/science/article/pii/S2215098615001664

254. Kato, S., Takeuchi, E., Ishiguro, Y., Ninomiya, Y., Takeda, K., Hamada, T.: An open approach to autonomous vehicles. IEEE Micro **35**(6), 60–68 (2015)

255. Kato, S., Tokunaga, S., Maruyama, Y., Maeda, S., Hirabayashi, M., Kitsukawa, Y., Monrroy, A., Ando, T., Fujii, Y., Azumi, T.: Autoware on board: Enabling autonomous vehicles with embedded systems. In: 2018 ACM/IEEE 9th International Conference on Cyber-Physical Systems (ICCPS), pp. 287–296. IEEE, New York (2018)

256. Ke, G., Meng, Q., Finley, T., Wang, T., Chen, W., Ma, W., Ye, Q., Liu, T.Y.: LightGBM: a highly efficient gradient boosting decision tree. In: Advances in Neural Information Processing Systems, pp. 3146–3154 (2017)

257. Kendall, A., Hawke, J., Janz, D., Mazur, P., Reda, D., Allen, J.M., Lam, V.D., Bewley, A., Shah, A.: Learning to drive in a day. In: 2019 International Conference on Robotics and Automation (ICRA), pp. 8248–8254. IEEE, New York (2019)

258. Kenney, J.B.: Dedicated short-range communications (DSRC) standards in the United States. Proc. IEEE **99**(7), 1162–1182 (2011)

259. Kesting, A., Treiber, M., Helbing, D.: General lane-changing model MOBIL for car-following models. Transport. Res. Rec. **1999**(1), 86–94 (2007)

260. Khan, U., Basaras, P., Schmidt-Thieme, L., Nanopoulos, A., Katsaros, D.: Analyzing cooperative lane change models for connected vehicles. In: 2014 International Conference on Connected Vehicles and Expo (ICCVE), pp. 565–570. IEEE, New York (2014)

261. Kim, B., Yim, J., Kim, J.: Highway driving dataset for semantic video segmentation (2020). Corpus ID: 52286240

262. Kim, J.H.: Estimating classification error rate: repeated cross-validation, repeated hold-out and bootstrap. Comput. Stat. Data Anal. **53**(11), 3735–3745 (2009)

263. Kim, J.H., Seo, S., Hai, N., Cheon, B.M., Lee, Y.S., Jeon, J.W.: Gateway framework for in-vehicle networks based on CAN, FlexRay, and ethernet. IEEE Trans. Veh. Technol. **64**(10), 4472–4486 (2015)

264. Kim, Z.: Robust lane detection and tracking in challenging scenarios. IEEE Trans. Intell. Transport. Syst. **9**(1), 16–26 (2008)

265. Ko, Y., Jun, J., Ko, D., Jeon, M.: Key points estimation and point instance segmentation approach for lane detection (2020). Preprint. arXiv:2002.06604

266. Kocić, J., Jovičić, N., Drndarević, V.: Sensors and sensor fusion in autonomous vehicles. In: 2018 26th Telecommunications Forum (TELFOR), pp. 420–425. IEEE, New York (2018)

267. Koenig, N., Howard, A.: Design and use paradigms for gazebo, an open-source multi-robot simulator. In: 2004 IEEE/RSJ International Conference on Intelligent Robots and Systems (IROS)(IEEE Cat. No. 04CH37566), vol. 3, pp. 2149–2154. IEEE, New York (2004)

268. Kohavi, R., et al.: A study of cross-validation and bootstrap for accuracy estimation and model selection. In: International Joint Conference on Artificial Intelligence (IJCAI), vol. 14, pp. 1137–1145 (1995)

269. Komkov, S., Petiushko, A.: AdvHat: Real-world adversarial attack on ArcFace face ID system (2019)

270. Konečný, J., McMahan, H.B., Yu, F.X., Richtárik, P., Suresh, A.T., Bacon, D.: Federated learning: strategies for improving communication efficiency (2016). Preprint. arXiv:1610.05492

271. Kong, J., Pfeiffer, M., Schildbach, G., Borrelli, F.: Kinematic and dynamic vehicle models for autonomous driving control design, pp. 1094–1099 (2015). https://doi.org/10.1109/IVS. 2015.7225830

272. Konolige, K.: Large-scale map-making. In: AAAI, pp. 457–463 (2004)

273. Konovaltsev, A., Cuntz, M., Hättich, C., Meurer, M.: Autonomous spoofing detection and mitigation in a GNSS receiver with an adaptive antenna array. In: ION GNSS+ 2013. The Institute of Navigation (2013). https://elib.dlr.de/86230/

274. Koopman, P., Wagner, M.: Challenges in autonomous vehicle testing and validation. SAE Int. J. Transport. Safe. **4**(1), 15–24 (2016)

275. Koscher, K., Czeskis, A., Roesner, F., Patel, S., Kohno, T., Checkoway, S., McCoy, D., Kantor, B., Anderson, D., Shacham, H., Savage, S.: Experimental security analysis of a modern automobile. In: 2010 IEEE Symposium on Security and Privacy, pp. 447–462 (2010)

276. Kotseruba, I., Rasouli, A., Tsotsos, J.K.: Joint attention in autonomous driving (JAAD) (2016). Preprint. arXiv:1609.04741

277. Krzanich, B.: Data is the new oil in the future of automated driving (2016). https://newsroom. intel.com/editorials/krzanich-the-future-of-automated-driving/

278. Kuutti, S., Fallah, S., Katsaros, K., Dianati, M., Mccullough, F., Mouzakitis, A.: A survey of the state-of-the-art localisation techniques and their potentials for autonomous vehicle applications. IEEE Int. Things J. **PP**, 1–1 (2018). https://doi.org/10.1109/JIOT.2018. 2812300

279. Labbé, M., Michaud, F.: RTAB-Map as an open-source LiDAR and visual simultaneous localization and mapping library for large-scale and long-term online operation: LabbÉ and michaud. J. Field Robot. **36** (2018). https://doi.org/10.1002/rob.21831

280. labelImg (2021). https://github.com/tzutalin/labelImg

281. LabelMe (2021). http://labelme.csail.mit.edu/Release3.0/

282. Lafuente-Arroyo, S., Gil-Jimenez, P., Maldonado-Bascon, R., López-Ferreras, F., Maldonado-Bascon, S.: Traffic sign shape classification evaluation I: SVM using distance to borders. In: IEEE Proceedings. Intelligent Vehicles Symposium, 2005, pp. 557–562. IEEE, New York (2005)

283. LaValle, S., Kuffner, J.: Randomized kinodynamic planning. pp. 473–479 (1999). https:// doi.org/10.1109/ROBOT.1999.770022

284. Leal-Taixé, L., Milan, A., Reid, I., Roth, S., Schindler, K.: MOTChallenge 2015: towards a benchmark for multi-target tracking (2015). Preprint. arXiv:1504.01942

285. Lee, J.W., Litkouhi, B.B.: System diagnosis in autonomous driving (2015). US Patent 9,168,924

286. Lee, K., Flinn, J., Noble, B.D.: Gremlin: scheduling interactions in vehicular computing. In: Proceedings of the 2nd ACM/IEEE Symposium on Edge Computing, pp. 1–13 (2017)

287. Lee, S., Kim, J., Shin Yoon, J., Shin, S., Bailo, O., Kim, N., Lee, T.H., Seok Hong, H., Han, S.H., So Kweon, I.: VPGNET: vanishing point guided network for lane and road marking detection and recognition. In: Proceedings of the IEEE International Conference on Computer Vision, pp. 1947–1955 (2017)

288. Lenchner, J., Isci, C., Kephart, J.O., Mansley, C., Connell, J., McIntosh, S.: Towards data center self-diagnosis using a mobile robot. In: Proceedings of the 8th ACM International Conference on Autonomic Computing, ICAC '11, pp. 81–90. ACM, New York (2011). http://doi.acm.org/10.1145/1998582.1998597

289. Leopard Imaging: Leopard Imaging AR023Z (2019). https://leopardimaging.com/product/li-usb30-ar023zwdr/

290. Leopard Imaging Inc.: USB 3.0 Box Camera: AR023ZWDRB (2018). https://leopardimaging.com/product/li-usb30-ar023zwdrb/

291. LeVine, S.: What it really costs to turn a car into a self-driving vehicle (2017). https://qz.com/924212/what-it-really-costs-to-turn-a-car-into-a-self-driving-vehicle/

292. Levinson, J., Askeland, J., Becker, J., Dolson, J., Held, D., Kammel, S., Kolter, J.Z., Langer, D., Pink, O., Pratt, V., et al.: Towards fully autonomous driving: Systems and algorithms. In: 2011 IEEE Intelligent Vehicles Symposium (IV), pp. 163–168. IEEE, New York (2011)

293. Levinson, J., Montemerlo, M., Thrun, S.: Map-based precision vehicle localization in Urban environments. In: Robotics: Science and Systems, vol. 4, p. 1. Citeseer (2007)

294. Li, B., Wang, W., Jia, L., Wang, D., Kong, A.: Study on HIL system of electric vehicle controller based on NI. In: IOP Conference Series: Materials Science and Engineering, vol. 382, p. 052033. IOP Publishing, Bristol (2018)

295. Li, B., Zhang, T., Xia, T.: Vehicle detection from 3d lidar using fully convolutional network (2016). Preprint. arXiv:1608.07916

296. Li, F., Li, Z., Han, W., Wu, T., Chen, L., Guo, Y., Chen, J.: Cyberspace-oriented access control: a cyberspace characteristics-based model and its policies. IEEE Internet Things J. 6(2), 1471–1483 (2019). https://doi.org/10.1109/JIOT.2018.2839065

297. Li, H., Ma, D., Medjahed, B., Kim, Y.S., Mitra, P.: Analyzing and preventing data privacy leakage in connected vehicle services. SAE Int. J. Adv. Curr. Prac. Mobil. 1, 1035–1045 (2019). https://doi.org/10.4271/2019-01-0478

298. Li, H.P., Li, Y.w.: The research of electric vehicle's MCU system based on iso26262. In: 2017 2nd Asia-Pacific Conference on Intelligent Robot Systems (ACIRS), pp. 336–340. IEEE, New York (2017)

299. Li, J., Mei, X., Prokhorov, D., Tao, D.: Deep neural network for structural prediction and lane detection in traffic scene. IEEE Trans. Neural Netw. Learn. Syst. 28(3), 690–703 (2016)

300. Li, S., Li, K., R.Rajamani, J.Wang: Model predictive multi-objective vehicular adaptive cruise control. IEEE Trans. Control Syst. Technol. 19(3), 556–566 (2011)

301. Li, X., Flohr, F., Yang, Y., Xiong, H., Braun, M., Pan, S., Li, K., Gavrila, D.M.: A new benchmark for vision-based cyclist detection. In: 2016 IEEE Intelligent Vehicles Symposium (IV), pp. 1028–1033. IEEE, New York (2016)

302. Lianos, K.N., Schönberger, J., Pollefeys, M., Sattler, T.: VSO: visual semantic odometry (2018)

303. Liaw, A., Wiener, M., et al.: Classification and regression by randomForest. R news 2(3), 18–22 (2002)

304. Lin, L., Liao, X., Jin, H., Li, P.: Computation offloading toward edge computing. Proc. IEEE 107, 1584–1607 (2019). https://doi.org/10.1109/JPROC.2019.2922285

305. Lin, S.C., Zhang, Y., Hsu, C.H., Skach, M., Haque, M.E., Tang, L., Mars, J.: The architectural implications of autonomous driving: Constraints and acceleration. In: ACM SIGPLAN Notices, vol. 53, pp. 751–766. ACM, New York (2018)

306. Lin, T.Y., Goyal, P., Girshick, R., He, K., Dollár, P.: Focal loss for dense object detection. In: Proceedings of the IEEE International Conference on Computer Vision, pp. 2980–2988 (2017)

307. Lin, T.Y., Maire, M., Belongie, S., Hays, J., Perona, P., Ramanan, D., Dollár, P., Zitnick, C.L.: Microsoft COCO: Common objects in context. In: European Conference on Computer Vision, pp. 740–755. Springer, New York (2014)

308. Liu, G., Wörgötter, F., Markelić, I.: Combining statistical hough transform and particle filter for robust lane detection and tracking. In: 2010 IEEE Intelligent Vehicles Symposium, pp. 993–997. IEEE, New York (2010)

309. Liu, L., Chen, J., Brocanelli, M., Shi, W.: E2M: an energy-efficient middleware for computer vision applications on autonomous mobile robots. In: Proceedings of the 4th ACM/IEEE Symposium on Edge Computing, pp. 59–73 (2019)

310. Liu, L., Lu, S., Zhong, R., Wu, B., Yao, Y., Zhang, Q., Shi, W.: Computing systems for autonomous driving: state-of-the-art and challenges. IEEE Internet Things J. (2020). arXiv:2009.14349v3 [cs.DC]

311. Liu, L., Wu, B., Suo, J., Shi, W.: Determinism analysis of deep neural network inference for autonomous driving. Technical report, CARTR-2020-12 (2020)

312. Liu, L., Yao, Y., Wang, R., Wu, B., Shi, W.: Equinox: a road-side edge computing experimental platform for CAVs. In: 2020 International Conference on Connected and Autonomous Driving (MetroCAD), pp. 41–42. IEEE, New York (2020)

313. Liu, L., Zhang, X., Qiao, M., Shi, W.: SafeShareRide: edge-based attack detection in ridesharing services. In: 2018 IEEE/ACM Symposium on Edge Computing (SEC), pp. 17–29 (2018). https://doi.org/10.1109/SEC.2018.00009

314. Liu, L., Zhang, X., Zhang, Q., Weinert, A., Wang, Y., Shi, W.: AutoVAPS: an IoT-enabled public safety service on vehicles. In: Proceedings of the Fourth Workshop on International Science of Smart City Operations and Platforms Engineering, SCOPE '19, pp. 41–47. ACM, New York (2019). http://doi.org/10.1145/3313237.3313303

315. Liu, P., Willis, D., Banerjee, S.: Paradrop: enabling lightweight multi-tenancy at the network's extreme edge. In: 2016 IEEE/ACM Symposium on Edge Computing (SEC), pp. 1–13. IEEE, New York (2016)

316. Liu, S., Li, L., Tang, J., Wu, S., Gaudiot, J.L.: Creating autonomous vehicle systems. Synth. Lect. Comput. Sci. 6(1), i–186 (2017)

317. Liu, S., Liu, L., Tang, J., Yu, B., Wang, Y., Shi, W.: Edge computing for autonomous driving: opportunities and challenges. Proc. IEEE 107(8), 1697–1716 (2019)

318. Liu, S., Tang, J., Wang, C., Wang, Q., Gaudiot, J.L.: A unified cloud platform for autonomous driving. Computer 50(12), 42–49 (2017)

319. Liu, S., Tang, J., Zhang, Z., Gaudiot, J.L.: Computer architectures for autonomous driving. Computer 50(8), 18–25 (2017). https://doi.org/10.1109/MC.2017.3001256

320. Liu, W., Anguelov, D., Erhan, D., Szegedy, C., Reed, S., Fu, C.Y., Berg, A.C.: SSD: single shot multibox detector. In: European Conference on Computer Vision, pp. 21–37. Springer, New York (2016)

321. Liu, X., Deng, Z., Lu, H., Cao, L.: Benchmark for road marking detection: dataset specification and performance baseline. In: ITSC (2017)

322. Liu, Y., Tight, M., Sun, Q., Kang, R.: A systematic review: road infrastructure requirement for Connected and Autonomous Vehicles (CAVs). J. Phys. Conf. Ser. 1187, 042073. IOP Publishing (2019)

323. Loose, H., Franke, U., Stiller, C.: Kalman particle filter for lane recognition on rural roads. In: 2009 IEEE Intelligent Vehicles Symposium, pp. 60–65. IEEE, New York (2009)

324. López, A., Serrat, J., Canero, C., Lumbreras, F., Graf, T.: Robust lane markings detection and road geometry computation. Int. J. Autom. Technol. 11(3), 395–407 (2010)

325. Lovejoy, B.: Apple moves to third-generation Siri back-end, built on open-source Mesos platform. (2015). https://9to5mac.com/2015/04/27/siri-backend-mesos/

326. Lu, S., Luo, B., Patel, T., Yao, Y., Tiwari, D., Shi, W.: Making disk failure predictions SMARTer! In: 18th USENIX Conference on File and Storage Technologies (FAST), pp. 151–167 (2020)

327. Lu, S., Yuan, X., Shi, W.: EdgeCompression: an integrated framework for compressive imaging processing on CAVs. In: Proceedings of the 5th ACM/IEEE Symposium on Edge Computing (SEC) (2020)

328. Luft, A.: The Chevrolet Sonic's days are numbered (2020). https://gmauthority.com/blog/2020/07/the-chevrolet-sonics-days-are-numbered/

329. Luo, P., Zhu, Z., Liu, Z., Wang, X., Tang, X.: Face model compression by distilling knowledge from neurons. In: Proceedings of the Thirtieth AAAI Conference on Artificial Intelligence, AAAI'16, pp. 3560–3566. AAAI Press, Cambridge (2016)

330. Luszczek, P.R., Bailey, D.H., Dongarra, J.J., Kepner, J., Lucas, R.F., Rabenseifner, R., Takahashi, D.: The HPC Challenge (HPCC) benchmark suite. In: Proceedings of the 2006 ACM/IEEE Conference on Supercomputing, p. 213. Citeseer (2006)

331. Lv, C., Hu, X., Sangiovanni-Vincentelli, A., Li, Y., Martinez, C.M., Cao, D.: Driving-style-based codesign optimization of an automated electric vehicle: a cyber-physical system approach. IEEE Trans. Ind. Electron. **66**(4), 2965–2975 (2018)

332. Lyu, Q., Qi, Y., Zhang, X., Liu, H., Wang, Q., Zheng, N.: SBAC: a secure blockchain-based access control framework for information-centric networking. J. Netw. Comput. Appl. **149**, 102444 (2020). https://doi.org/10.1016/j.jnca.2019.102444

333. Ma, C., Dai, X., Zhu, J., Liu, N., Sun, H., Liu, M.: Drivingsense: dangerous driving behavior identification based on smartphone autocalibration. Mob. Inf. Syst. **2017** (2017). https://doi.org/10.1155/2017/9075653

334. Ma, L., Yi, S., Li, Q.: Efficient service handoff across edge servers via docker container migration. In: Proceedings of the Second ACM/IEEE Symposium on Edge Computing, pp. 1–13 (2017)

335. Ma, Y., Zhang, K., Gu, J., Li, J., Lu, D.: Design of the control system for a four-wheel driven micro electric vehicle. In: 2009 IEEE Vehicle Power and Propulsion Conference, pp. 1813–1816. IEEE, New York (2009)

336. Maas, A.L., Hannun, A.Y., Lengerich, C.T., Qi, P., Jurafsky, D., Ng, A.Y.: Increasing deep neural network acoustic model size for large vocabulary continuous speech recognition (2014). arXiv. Preprint

337. Maciej, J., Vollrath, M.: Comparison of manual vs. speech-based interaction with in-vehicle information systems. Accid. Anal. Prev. **41**(5), 924–930 (2009)

338. Maddern, W., Pascoe, G., Linegar, C., Newman, P.: 1 year, 1000 km: the oxford robotcar dataset. Int. J. Robot. Res. **36**(1), 3–15 (2017)

339. Maeda, H., Sekimoto, Y., Seto, T., Kashiyama, T., Omata, H.: Road damage detection using deep neural networks with images captured through a smartphone (2018). Preprint. arXiv:1801.09454

340. Maene, P., Götzfried, J., de Clercq, R., Müller, T., Freiling, F., Verbauwhede, I.: Hardware-based trusted computing architectures for isolation and attestation. IEEE Trans. Comput. **67**(3), 361–374 (2018). https://doi.org/10.1109/TC.2017.2647955

341. Makesense.ai (2021). https://www.makesense.ai/

342. Magnusson, M., Nuchter, A., Lorken, C., Lilienthal, A., Hertzberg, J.: Evaluation of 3D registration reliability and speed - a comparison of ICP and NDT, pp. 3907–3912 (2009). https://doi.org/10.1109/ROBOT.2009.5152538

343. Mao, Y., Yi, S., Li, Q., Feng, J., Xu, F., Zhong, S.: Learning from differentially private neural activations with edge computing. In: 2018 IEEE/ACM Symposium on Edge Computing (SEC), pp. 90–102. IEEE, New York (2018)

344. Marjovi, A., Vasic, M., Lemaitre, J., Martinoli, A.: Distributed graph-based convoy control for networked intelligent vehicles. In: Intelligent Vehicles Symposium (IV), 2015 IEEE, pp. 138–143. IEEE, New York (2015)

345. Martinelli, F., Mercaldo, F., Orlando, A., Nardone, V., Santone, A., Sangaiah, A.K.: Human behavior characterization for driving style recognition in vehicle system. Comput. Electr. Eng. **83**, 102504 (2020). https://doi.org/10.1016/j.compeleceng.2017.12.050. http://www.sciencedirect.com/science/article/pii/S0045790617329531

346. Martinez, J., Canudas-De-Wit, C.: A safe longitudinal control for adaptive cruise control and stop-and-go scenarios. IEEE Trans. Control Syst. Technol. **15**(2), 246–258 (2007)

347. McLurkin, J., McMullen, A., Robbins, N., Habibi, G., Becker, A., Chou, A., Li, H., John, M., Okeke, N., Rykowski, J., et al.: A robot system design for low-cost multi-robot manipulation. In: 2014 IEEE/RSJ International Conference on Intelligent Robots and Systems, pp. 912–918. IEEE, New York (2014)

348. Mcmanus, C., Churchill, W., Napier, A., Davis, B., Newman, P.: Distraciton suppression for vision-based pose estimation at city scales (2013). https://doi.org/10.1109/ICRA.2013.6631106

349. Mearian, B.L.: Self-driving cars could create 1GB of data a second (2013). https://www.computerworld.com/article/2484219/emerging-technology/self-driving-cars-could-create-1gb-of-data-a-second.html

350. Meet NVIDIA Xavier: A new brain for self-driving, AI, and AR cars (2018). https://www.slashgear.com/meet-nvidia-xavier-a-new-brain-for-self-driving-ai-and-ar-cars-07513987/

351. Meet the cruise av: the first production-ready car with no steering wheel or pedals (2018). http://media.gm.com/media/us/en/gm/home.detail.html/content/Pages/news/us/en/2018/jan/0112-cruise-av.html

352. Mei, Y., Lu, Y.H., Hu, Y.C., Lee, C.S.G.: Energy-efficient motion planning for mobile robots. In: IEEE International Conference on Robotics and Automation, 2004. Proceedings. ICRA '04. 2004, vol. 5, pp. 4344–4349 (2004). https://doi.org/10.1109/ROBOT.2004.1302401

353. Miovision unveils the world's smartest intersection in detroit (2018). https://miovision.com/press/miovision-unveils-the-worlds-smartest-intersection-in-detroit/

354. Meyer, M., Kuschk, G.: Automotive radar dataset for deep learning based 3d object detection. In: 2019 16th European Radar Conference (EuRAD), pp. 129–132. IEEE, New York (2019)

355. Michaelis, C., Mitzkus, B., Geirhos, R., Rusak, E., Bringmann, O., Ecker, A.S., Bethge, M., Brendel, W.: Benchmarking robustness in object detection: autonomous driving when winter is coming (2019). Preprint. arXiv:1907.07484

356. Milan, A., Leal-Taixé, L., Reid, I., Roth, S., Schindler, K.: Mot16: a benchmark for multi-object tracking (2016). Preprint. arXiv:1603.00831

357. Mills, D.: RFC1305: Network Time Protocol (Version 3) Specification, Implementation. RFC Editor (1992)

358. Mofrad, S., Zhang, F., Lu, S., Shi, W.: A comparison study of intel SGX and AMD memory encryption technology. In: Proceedings of The Hardware and Architectural Support for Security and Privacy (HSAP'18) (2018). https://doi.org/10.1145/3214292.3214301

359. Mogren, O.: C-RNN-GAN: continuous recurrent neural networks with adversarial training (2016). Preprint. arXiv:1611.09904

360. MongoDB, Inc: The most popular database for modern apps | MongoDB. https://www.mongodb.com/ (2019). Accessed 30 Sept 2019

361. Monjazeb, A., Sasiadek, J.Z., Necsulescu, D.: Autonomous navigation among large number of nearby landmarks using FastSLAM and EKF-SLAM-A comparative study. In: 2011 16th International Conference on Methods & Models in Automation & Robotics, pp. 369–374. IEEE, New York (2011)

362. Montemerlo, M., Becker, J., Bhat, S., Dahlkamp, H., Dolgov, D., Ettinger, S., Haehnel, D., Hilden, T., Hoffmann, G., Huhnke, B., et al.: Junior: The Stanford entry in the urban challenge. J. Field Robot. **25**(9), 569–597 (2008)

363. Morsink, P., Hallouzi, R., Dagli, I., Cseh, C., Schäfers, L., Nelisse, M., Bruin, D.D.: CarTALK 2000: development of a co-operative ADAS based on vehicle-to-vehicle communication. In: 10th World Congress and Exhibition on Intelligent Transport Systems and Services, 16-20 November, 2003, Madrid (2003)

364. Mortazavi, S.H., Balasubramanian, B., de Lara, E., Narayanan, S.P.: Toward session consistency for the edge. In: USENIX Workshop on Hot Topics in Edge Computing (HotEdge 18) (2018)

365. Mousavian, A., Anguelov, D., Flynn, J., Kosecka, J.: 3d bounding box estimation using deep learning and geometry (2017)

366. Mozilla: Mozilla Corpus (2018). https://voice.mozilla.org/en/data

367. Mullane, J., Vo, B.N., Adams, M.D., Wijesoma, W.S.: A random set formulation for Bayesian slam. In: 2008 IEEE/RSJ International Conference on Intelligent Robots and Systems, pp. 1043–1049. IEEE, New York (2008)

368. Muller, U., Ben, J., Cosatto, E., Flepp, B., Cun, Y.L.: Off-road obstacle avoidance through end-to-end learning. In: Advances in Neural Information Processing Systems, pp. 739–746 (2006)

369. Mur-Artal, R., Montiel, J.M.M., Tardos, J.D.: ORB-SLAM: a versatile and accurate monocular SLAM system. IEEE Trans. Robot. **31**(5), 1147–1163 (2015)
370. Mur Atal, R., Tardos, J.D.: ORB-SLAM2: an Open-Source SLAM system for monocular, stereo, and RGB-D camera. IEEE Trans. Robot. **33**(5), 1255–1262 (2017). https://doi.org/10.1109/MPRV.2009.82
371. Mur-Artal, R., Tardós, J.D.: ORB-SLAM2: an open-source SLAM system for monocular, stereo, and RGB-D cameras. IEEE Trans. Robot. **33**(5), 1255–1262 (2017)
372. Mysql (2018). https://www.mysql.com/
373. Naic: Usage-based insurance and telematics, https://www.naic.org (2018). https://www.naic.org/cipr_topics/topic_usage_based_insurance.htm
374. Nanda, A., Puthal, D., Rodrigues, J.J.P.C., Kozlov, S.A.: Internet of autonomous vehicles communications security: overview, issues, and directions. IEEE Wirel. Commun. **26**(4), 60–65 (2019)
375. Nardi, L., Bodin, B., Zia, M.Z., Mawer, J., Nisbet, A., Kelly, P.H., Davison, A.J., Luján, M., O'Boyle, M.F., Riley, G., et al.: Introducing SLAMBench, a performance and accuracy benchmarking methodology for slam. In: 2015 IEEE International Conference on Robotics and Automation (ICRA), pp. 5783–5790. IEEE, New York (2015)
376. NATS: Nats - open source messaging system. https://nats.io/ (2019). Accessed 25 Sept 2019
377. Nellans, D., Nellans, D., Bonnet, P.: Linux block IO: introducing multi-queue ssd access on multi-core systems. In: International Systems and Storage Conference, p. 22 (2013)
378. Nelson, P.: Just one autonomous car will use 4,000 gb of data/day (2016). http://www.networkworld.com/article/3147892/internet/one-autonomous-car-will-use-4000-gb-of-dataday.html
379. Neuhold, G., Ollmann, T., Rota Bulo, S., Kontschieder, P.: The mapillary vistas dataset for semantic understanding of street scenes. In: Proceedings of the IEEE International Conference on Computer Vision, pp. 4990–4999 (2017)
380. Neven, D., De Brabandere, B., Georgoulis, S., Proesmans, M., Van Gool, L.: Towards end-to-end lane detection: an instance segmentation approach. In: 2018 IEEE Intelligent Vehicles Symposium (IV), pp. 286–291. IEEE, New York (2018)
381. New autonomous mileage reports are out, but is the data meaningful? http://bit.ly/AMRData (2019). Online. Accessed 13 Dec 2020
382. Newcombe, R., Lovegrove, S., Davison, A.: DTAM: dense tracking and mapping in real-time. pp. 2320–2327 (2011). https://doi.org/10.1109/ICCV.2011.6126513
383. Newcombe, R.A., Izadi, S., Hilliges, O., Molyneaux, D., Kim, D., Davison, A.J., Kohi, P., Shotton, J., Hodges, S., Fitzgibbon, A.: KinectFusion: real-time dense surface mapping and tracking. In: 2011 10th IEEE international symposium on Mixed and Augmented Reality (ISMAR), pp. 127–136. IEEE, New York (2011)
384. NGINX Inc: Nginx | high performance load balancer, web server, and reverse proxy (2019). https://www.nginx.com/. Accessed 15 Sept 2019
385. Nikaein, N., Marina, M.K., Manickam, S., Dawson, A., Knopp, R., Bonnet, C.: OpenAir-Interface: a flexible platform for 5G research. ACM SIGCOMM Comput. Commun. Rev. **44**(5), 33–38 (2014)
386. Nilsson, D.K., Larson, U.E., Picasso, F., Jonsson, E.: A first simulation of attacks in the automotive network communications protocol flexray. In: Corchado, E., Zunino, R., Gastaldo, P., Herrero, Á. (eds.) Proceedings of the International Workshop on Computational Intelligence in Security for Information Systems CISIS'08, pp. 84–91. Springer, Berlin Heidelberg (2009)
387. Ning, Z., Zhang, F., Shi, W., Shi, W.: Position paper: challenges towards securing hardware-assisted execution environments. In: Proceedings of the Hardware and Architectural Support for Security and Privacy, HASP '17. Association for Computing Machinery, New York (2017). https://doi.org/10.1145/3092627.3092633
388. Nurmi, D., Wolski, R., Grzegorczyk, C., Obertelli, G., Soman, S., Youseff, L., Zagorodnov, D.: The eucalyptus open-source cloud-computing system. In: 9th IEEE/ACM International Symposium on Cluster Computing and the Grid, 2009. CCGRID'09, pp. 124–131. IEEE, New York (2009)

389. NVIDIA Corporation: NVIDIA DRIVE PX2: Scalable AI platform for Autonomous Driving (2018). https://www.nvidia.com/en-us/self-driving-cars/drive-platform

390. NVIDIA DRIVE - Software (2020). https://developer.nvidia.com/drive/drive-software

391. NVIDIA Corporation: NVIDIA Jetson Platform (2019). https://www.nvidia.com/en-us/autonomous-machines/embedded-systems

392. O'Hanlon, B.W., Psiaki, M.L., Bhatti, J.A., Shepard, D.P., Humphreys, T.E.: Real-time GPS spoofing detection via correlation of encrypted signals. NAVIGATION **60**(4), 267–278 (2013). https://onlinelibrary.wiley.com/doi/abs/10.1002/navi.44

393. Olafenwa, M.: Traffic-net dataset (2021). https://github.com/OlafenwaMoses/Traffic-Net. Accessed 19 Jan 2021

394. OpenALPR Technology Inc.: OpenALPR (2018). http://www.openalpr.com

395. Asam opendrive®. http://bit.ly/ASAMOpenDrive. Online. Accessed 30 Dec 2020

396. Organization, W.H.: Global Status Report on Road Safety 2015. World Health Organization (2015)

397. Orrick, A., McDermott, M., Barnett, D.M., Nelson, E.L., Williams, G.N.: Failure detection in an autonomous underwater vehicle. In: Proceedings of IEEE Symposium on Autonomous Underwater Vehicle Technology (AUV'94), pp. 377–382. IEEE, New York (1994)

398. Osep, A., Mehner, W., Mathias, M., Leibe, B.: Combined image-and world-space tracking in traffic scenes. In: 2017 IEEE International Conference on Robotics and Automation (ICRA), pp. 1988–1995. IEEE, New York (2017)

399. Pala, Z., Inanc, N.: Smart parking applications using RFID technology. pp. 1–3 (2007). https://doi.org/10.1109/RFIDEURASIA.2007.4368108

400. Pan, S.J., Yang, Q.: A survey on transfer learning. IEEE Trans. Knowl. Data Eng. **22**(10), 1345–1359 (2010)

401. Pan, X., Shi, J., Luo, P., Wang, X., Tang, X.: Spatial as deep: spatial CNN for traffic scene understanding. In: Thirty-Second AAAI Conference on Artificial Intelligence (2018)

402. Pandey, G., McBride, J.R., Eustice, R.M.: Ford campus vision and lidar data set. Int. J. Robot. Res. **30**(13), 1543–1552 (2011)

403. Papathanassiou, A., Khoryaev, A.: Cellular V2X as the essential enabler of superior global connected transportation services. IEEE 5G Tech Focus **1**(2), 1–2 (2017)

404. Park, C.Y.: Predicting deterministic execution times of real-time programs (1992)

405. Park, D., Kim, S., An, Y., Jung, J.Y.: Lired: A light-weight real-time fault detection system for edge computing using LSTM recurrent neural networks. Sensors **18**(7), 2110 (2018)

406. Patil, A., Malla, S., Gang, H., Chen, Y.T.: The H3D dataset for full-surround 3d multi-object detection and tracking in crowded urban scenes. In: 2019 International Conference on Robotics and Automation (ICRA) (2019)

407. Pedregosa, F., Varoquaux, G., Gramfort, A., Michel, V., Thirion, B., Grisel, O., Blondel, M., Prettenhofer, P., Weiss, R., Dubourg, V., et al.: Scikit-learn: machine learning in python. J. Mach. Learn. Res. **12**, 2825–2830 (2011)

408. Pekhimenko, G., Guo, C., Jeon, M., Huang, P., Zhou, L.: Tersecades: efficient data compression in stream processing. In: Proceedings of the 2018 USENIX Conference on Usenix Annual Technical Conference, USENIX ATC '18, p. 307–320. USENIX Association, USA (2018)

409. Peled, E., Golodnitsky, D., Mazor, H., Goor, M., Avshalomov, S.: Parameter analysis of a practical lithium-and sodium-air electric vehicle battery. J. Power Sources **196**(16), 6835–6840 (2011)

410. Peng, X., Hou, J., Tan, L., Chen, J., Jiang, J., Guo, X.: Bit-Error aware lossless color image compression. In: 2019 IEEE International Conference on Electro Information Technology (EIT), pp. 126–131 (2019). https://doi.org/10.1109/EIT.2019.8833786

411. PerceptIn's self-driving vehicles go on sale in November for $40,000 (2018). https://venturebeat.com/2018/09/12/perceptins-self-driving-vehicles-go-on-sale-in-november-for-40000/

412. Petit, J., Stottelaar, B., Feiri, M., Kargl, F.: Remote attacks on automated vehicles sensors: experiments on camera and LiDAR. Black Hat Europe **11**, 2015 (2015)

413. Pham, Q.H., Uy, M.A., Hua, B.S., Nguyen, D.T., Roig, G., Yeung, S.K.: LCD: learned cross-domain descriptors for 2D-3D matching. Proceedings of the AAAI Conference on Artificial Intelligence **34**, 11856–11864 (2020). https://doi.org/10.1609/aaai.v34i07.6859

414. Philion, J.: FastDraw: addressing the long tail of lane detection by adapting a sequential prediction network. In: Proceedings of the IEEE Conference on Computer Vision and Pattern Recognition, pp. 11582–11591 (2019)

415. Pinggera, P., Ramos, S., Gehrig, S., Franke, U., Rother, C., Mester, R.: Lost and found: detecting small road hazards for self-driving vehicles. In: 2016 IEEE/RSJ International Conference on Intelligent Robots and Systems (IROS) (2016)

416. Pitropov, M., Garcia, D., Rebello, J., Smart, M., Wang, C., Czarnecki, K., Waslander, S.: Canadian adverse driving conditions dataset. Int. J. Robot. Res. **40**(4–5), 681–690 (2020)

417. Polastre, J., Szewczyk, R., Culler, D.: Telos: enabling ultra-low power wireless research. In: Proceedings of the 4th International Symposium on Information Processing in Sensor Networks, p. 48. IEEE Press, New York (2005)

418. Popa, R.A., Redfield, C.M.S., Zeldovich, N., Balakrishnan, H.: CryptDB: protecting confidentiality with encrypted query processing. In: Proceedings of the Twenty-Third ACM Symposium on Operating Systems Principles, SOSP '11, p. 85–100. Association for Computing Machinery, New York (2011). https://doi.org/10.1145/2043556.2043566

419. Prexl, M., Zunhammer, N., Walter, U.: Motion prediction for teleoperating autonomous vehicles using a PID control model, pp. 133–138 (2019). https://doi.org/10.1109/ANZCC47194.2019.8945623

420. Qi, B., Kang, L., Banerjee, S.: A vehicle-based edge computing platform for transit and human mobility analytics. In: Proceedings of the Second ACM/IEEE Symposium on Edge Computing, San Jose/Silicon Valley, SEC 2017, CA, October 12–14, 2017, pp. 1:1–1:14 (2017). http://doi.acm.org/10.1145/3132211.3134446

421. Qi, B., Kang, L., Banerjee, S.: A vehicle-based edge computing platform for transit and human mobility analytics. In: Proceedings of the Second ACM/IEEE Symposium on Edge Computing, p. 1. ACM, New York (2017)

422. Qi, H., Feng, C., Cao, Z., Zhao, F., Xiao, Y.: P2b: Point-to-box network for 3d object tracking in point clouds. In: Proceedings of the IEEE/CVF Conference on Computer Vision and Pattern Recognition, pp. 6329–6338 (2020)

423. Qu, F., Wu, Z., Wang, F., Cho, W.: A security and privacy review of VANETs. IEEE Trans. Intell. Transport. Syst. **16**(6), 2985–2996 (2015)

424. Quigley, M., Conley, K., Gerkey, B., Faust, J., Foote, T., Leibs, J., Wheeler, R., Ng, A.Y.: ROS: an open-source robot operating system. In: ICRA Workshop on Open Source Software, Kobe, vol. 3, p. 5 (2009)

425. Qureshi, M.A., Deriche, M.: A new wavelet based efficient image compression algorithm using compressive sensing. Multimedia Tools Appl. **75**(12), 6737–6754 (2016). https://doi.org/10.1007/s11042-015-2590-9

426. Ramanishka, V., Chen, Y.T., Misu, T., Saenko, K.: Toward driving scene understanding: a dataset for learning driver behavior and causal reasoning. In: Proceedings of the IEEE Conference on Computer Vision and Pattern Recognition, pp. 7699–7707 (2018)

427. Ran, L., Junfeng, W., Haiying, W., Gechen, L.: Design method of can bus network communication structure for electric vehicle. In: International Forum on Strategic Technology 2010, pp. 326–329. IEEE, New York (2010)

428. Rao, Q., Frtunikj, J.: Deep learning for self-driving cars: chances and challenges. In: 2018 IEEE/ACM 1st International Workshop on Software Engineering for AI in Autonomous Systems (SEFAIAS), pp. 35–38 (2018)

429. Ravidas, S., Lekidis, A., Paci, F., Zannone, N.: Access control in internet-of-things: a survey. J. Netw. Comput. Appl. **144**, 79–101 (2019). https://doi.org/10.1016/j.jnca.2019.06.017

430. Rebecq, H., Horstschaefer, T., Gallego, G., Scaramuzza, D.: EVO: a geometric approach to event-based 6-DOF parallel tracking and mapping in real-time. IEEE Robot. Autom. Lett. **PP** (2016). https://doi.org/10.1109/LRA.2016.2645143

431. Rebecq, H., Ranftl, R., Koltun, V., Scaramuzza, D.: High speed and high dynamic range video with an event camera. In: IEEE Transactions on Pattern Analysis and Machine Intelligence (2019)
432. RectLabel (2021). https://rectlabel.com/
433. Redis (2018). https://redis.io/
434. Redmon, J., Divvala, S., Girshick, R., Farhadi, A.: You only look once: unified, real-time object detection. In: Proceedings of the IEEE Conference on Computer Vision and Pattern Recognition, pp. 779–788 (2016)
435. Redmon, J., Farhadi, A.: YOLO9000: better, faster, stronger. In: Proceedings of the IEEE Conference on Computer Vision and Pattern Recognition, pp. 7263–7271 (2017)
436. Redmon, J., Farhadi, A.: YOLOv3: An incremental improvement (2018). Preprint. arXiv:1804.02767
437. Regulation, P.: General data protection regulation. Off. J. Eur. Union **59**, 1–88 (2016)
438. Reichardt, D., Miglietta, M., Moretti, L., Morsink, P., Schulz, W.: CarTALK 2000: safe and comfortable driving based upon inter-vehicle-communication. In: Intelligent Vehicle Symposium, 2002. IEEE, vol. 2, pp. 545–550. IEEE, New York (2002)
439. Releases, U.: Fatal traffic crash data. Article (CrossRef Link) (2016)
440. Ren, K., Wang, Q., Wang, C., Qin, Z., Lin, X.: The security of autonomous driving: threats, defenses, and future directions. Proc. IEEE **108**(2), 357–372 (2020)
441. Ren, S., He, K., Girshick, R., Sun, J.: Faster R-CNN: Towards real-time object detection with region proposal networks. In: Advances in Neural Information Processing Systems, pp. 91–99 (2015)
442. Rippel, O., Bourdev, L.: Real-time adaptive image compression. In: Proceedings of the 34th International Conference on Machine Learning - Volume 70, ICML'17, pp. 2922–2930. JMLR.org (2017). http://dl.acm.org/citation.cfm?id=3305890.3305983
443. Robotic operating system (2020). https://www.ros.org/. Accessed 18 Feb 2020
444. Rodriguez, J.D., Perez, A., Lozano, J.A.: Sensitivity analysis of k-fold cross validation in prediction error estimation. IEEE Trans. Patt. Anal. Mach. Intell. **32**(3), 569–575 (2010)
445. Romera, E., Bergasa, L.M., Arroyo, R.: Need data for driver behaviour analysis? Presenting the public UAH-driveset. In: 2016 IEEE 19th International Conference on Intelligent Transportation Systems (ITSC), pp. 387–392. IEEE, New York (2016)
446. Rong, G., Shin, B.H., Tabatabaee, H., Lu, Q., Lemke, S., Možeiko, M., Boise, E., Uhm, G., Gerow, M., Mehta, S., et al.: lGSVL simulator: a high fidelity simulator for autonomous driving (2020). Preprint. arXiv:2005.03778
447. Ros, G., Sellart, L., Materzynska, J., Vazquez, D., Lopez, A.M.: The synthia dataset: a large collection of synthetic images for semantic segmentation of urban scenes. In: Proceedings of the IEEE Conference on Computer Vision and Pattern Recognition, pp. 3234–3243 (2016)
448. ROS 2 Documentation (2020). https://index.ros.org/doc/ros2/
449. Rossi, T.: Autonomous and adas test cars produce over 11 tb of data per day (article) (October 10, 2018). https://www.tuxera.com/blog/autonomous-and-adas-test-cars-produce-over-11-tb-of-data-per-day/
450. Roy, S., Das, A.K., Chatterjee, S., Kumar, N., Chattopadhyay, S., Rodrigues, J.J.P.C.: Provably secure fine-grained data access control over multiple cloud servers in mobile cloud computing based healthcare applications. IEEE Trans. Ind. Inf. **15**(1), 457–468 (2019). https://doi.org/10.1109/TII.2018.2824815
451. Ruijun Wang, L.L., Shi, W.: HydraSpace: computational data storage for autonomous vehicles. In: IEEE Collaborative and Internet Computing Vision Track (CIC), December (2020)
452. Ryu, J.H., Ogay, D., Bulavintsev, S., Kim, H., Park, J.S.: Development and experiences of an autonomous vehicle for high-speed navigation and obstacle avoidance. Front. Intell. Auton. Syst. **466**, 105–116 (2013). https://doi.org/10.1007/978-3-642-35485-4_8
453. Sabaliauskaite, G., Liew, L.S., Cui, J.: Integrating autonomous vehicle safety and security analysis using STPA method and the six-step model. Int. J. Adv. Secur. **11**(1&2), 160–169 (2018)

454. SAE International.: Taxonomy and definitions for terms related to driving automation systems for on-road motor vehicles J3016 (2016). https://www.sae.org/standards/content/j3016_201609/

455. SafeDI scenario-based AV policy framework – an overview for policy-makers (2020). https://www.weforum.org/whitepapers/safe-drive-initiative-safedi-scenario-based-av-policy-framework-an-overview-for-policy-makers

456. Sallab, A.E., Abdou, M., Perot, E., Yogamani, S.: Deep reinforcement learning framework for autonomous driving. Electron. Imag. **2017**(19), 70–76 (2017)

457. Sandhu, R.S., Samarati, P.: Access control: principle and practice. IEEE Commun. Mag. **32**(9), 40–48 (1994). https://doi.org/10.1109/35.312842

458. Santana, E., Hotz, G.: Learning a driving simulator (2016). Preprint. arXiv:1608.01230

459. Sarrafan, K., Muttaqi, K.M., Sutanto, D.: Real-time state-of-charge tracking system using mixed estimation algorithm for electric vehicle battery system. In: 2018 IEEE Industry Applications Society Annual Meeting (IAS), pp. 1–8. IEEE, New York (2018)

460. Sato, H., Yakoh, T.: A real-time communication mechanism for RTLinux. In: 2000 26th Annual Conference of the IEEE Industrial Electronics Society. IECON 2000. 2000 IEEE International Conference on Industrial Electronics, Control and Instrumentation. 21st Century Technologies, vol. 4, pp. 2437–2442. IEEE, New York (2000)

461. Satyanarayanan, M.: The emergence of edge computing. Computer **50**(1), 30–39 (2017). https://doi.org/10.1109/MC.2017.9

462. Satyanarayanan, M., Bahl, P., Caceres, R., Davies, N.: The case for VM-based cloudlets in mobile computing. IEEE Pervas. Comput. **8**(4), 14–23 (2009). https://doi.org/10.1109/MPRV.2009.82

463. Satyanarayanan, M., Simoens, P., Xiao, Y., Pillai, P., Chen, Z., Ha, K., Hu, W., Amos, B.: Edge analytics in the internet of things. IEEE Pervas. Comput. **14**(2), 24–31 (2015). https://doi.org/10.1109/MPRV.2015.32

464. Sau, B.B., Balasubramanian, V.N.: Deep model compression: distilling knowledge from noisy teachers. CoRR **abs/1610.09650** (2016). http://arxiv.org/abs/1610.09650

465. Sau, B.B., Balasubramanian, V.N.: Deep model compression: distilling knowledge from noisy teachers (2016). Preprint. arXiv:1610.09650

466. Savaglio, C., Gerace, P., Di Fatta, G., Fortino, G.: Data mining at the IoT edge. In: 2019 28th International Conference on Computer Communication and Networks (ICCCN), pp. 1–6. IEEE, New York (2019)

467. Scalable (2021). https://scalabel.ai/

468. Scale, H..: Pandaset. [EB/OL]. https://scale.com/open-datasets/pandaset. Accessed 24 Jan 2021

469. Schantz, R.E., Schmidt, D.C.: Middleware. In: Encyclopedia of Software Engineering. Wiley, Chichester (2002)

470. Schlegel, D., Colosi, M., Grisetti, G.: ProSLAM: graph SLAM from a programmer's perspective. In: 2018 IEEE International Conference on Robotics and Automation (ICRA), pp. 1–9 (2018). https://doi.org/10.1109/ICRA.2018.8461180

471. Schoettle, B.: Sensor fusion: a comparison of sensing capabilities of human drivers and highly automated vehicles. University of Michigan (2017). Corpus ID: 216011954

472. Schöner, H.: The role of simulation in development and testing of autonomous vehicles. In: Driving Simulation Conference, Stuttgart (2017)

473. Schulze, M., Nocker, G., Bohm, K.: PReVENT: A European program to improve active safety. In: Proceedings of 5th International Conference on Intelligent Transportation Systems Telecommunications, France (2005)

474. SDF. http://sdformat.org/. Online. Accessed 05 Dec 2020

475. SecurityInfoWatch: Data generated by new surveillance cameras to increase exponentially in the coming years (January 20, 2016). https://www.securityinfowatch.com/video-surveillance/news/12160483

476. Sedgwick, D.: When driverless cars call for backup (2017). https://www.autonews.com/article/20170218/OEM10/302209969/when-driverless-cars-call-for-backup

477. Sefraoui, O., Aissaoui, M., Eleuldj, M.: Openstack: toward an open-source solution for cloud computing. Int. J. Comput. Appl. **55**(3), 38–42 (2012)

478. Self-driving car (2019). https://en.wikipedia.org/wiki/Self-driving_car

479. Seo, Y.W., Lee, J., Zhang, W., Wettergreen, D.: Recognition of highway workzones for reliable autonomous driving. IEEE Trans. Intell. Transport. Syst. **16**, 1–11 (2014). https://doi.org/10.1109/TITS.2014.2335535

480. Serving DNNs like clockwork: performance predictability from the bottom up. In: 14th USENIX Symposium on Operating Systems Design and Implementation (OSDI 20). USENIX Association, Banff, Alberta (2020). https://www.usenix.org/conference/osdi20/presentation/gujarati

481. Shebaro, B., Oluwatimi, O., Bertino, E.: Context-based access control systems for mobile devices. IEEE Trans. Depend. Sec. Comput. **12**(2), 150–163 (2015). https://doi.org/10.1109/TDSC.2014.2320731

482. Shen, Z., Dai, Y., Rao, Z.: MSMD-Net: deep stereo matching with multi-scale and multi-dimension cost volume (2020). CoRR abs/2006.12797

483. Shi, W., Cao, J., Zhang, Q., Li, Y., Xu, L.: Edge computing: vision and challenges. IEEE Internet Things J. **3**(5), 637–646 (2016)

484. Shi, W., Dustdar, S.: The promise of edge computing. Computer **49**(5), 78–81 (2016)

485. Shin, H., Kim, D., Kwon, Y., Kim, Y.: Illusion and dazzle: adversarial optical channel exploits against LiDARs for automotive applications. In: Fischer, W., Homma, N. (eds.) Cryptographic Hardware and Embedded Systems – CHES 2017, pp. 445–467. Springer International Publishing, Cham (2017)

486. Shusterman, E., Feder, M., Member, S.: Image compression via improved quadtree decomposition algorithms. In: IEEE Transactions on Image Processing: A Publication of the IEEE Signal Processing Society (1994)

487. Slamtec: RPLIDAR A2 Laser Range Scanner (2019). https://www.slamtec.com/en/Lidar/A2

488. SMP Robotics: Application of autonomous mobile robots (2019). https://smprobotics.com/application_autonomus_mobile_robots/

489. Sochor, J., Špaňhel, J., Herout, A.: Boxcars: improving fine-grained recognition of vehicles using 3-d bounding boxes in traffic surveillance. IEEE Trans. Intell. Transport. Syst. **20**(1), 97–108 (2018)

490. Somerville, H., Lienert, P., Sage, A.: Uber's use of fewer safety sensors prompts questions after Arizona crash. Business news, Reuters (2018)

491. Standard, S.: J3016: Taxonomy and definitions for terms related to on-road motor vehicle automated driving systems (2014)

492. Standard, S.S.V.: Dedicated Short Range Communications (DSRC) message set dictionary. SAE International, November (2009)

493. Steinhauser, D., Ruepp, O., Burschka, D.: Motion segmentation and scene classification from 3d LiDAR data. In: 2008 IEEE Intelligent Vehicles Symposium, pp. 398–403 (2008). https://doi.org/10.1109/IVS.2008.4621281

494. Steve LeVine: What it really costs to turn a car into a self-driving vehicle (2017). https://qz.com/924212/what-it-really-costs-to-turn-a-car-into-a-self-driving-vehicle/

495. Stinson, D.R., Paterson, M.: Cryptography: Theory and Practice. CRC Press, Boca Raton (2018)

496. Stübing, H., Bechler, M., Heussner, D., May, T., Radusch, I., Rechner, H., Vogel, P.: SimTD: A car-to-x system architecture for field operational tests [topics in automotive networking]. IEEE Commun. Mag. **48**(5) (2010)

497. Sturm, J., Engelhard, N., Endres, F., Burgard, W., Cremers, D.: A benchmark for the evaluation of RGB-D SLAM systems. In: 2012 IEEE/RSJ International Conference on Intelligent Robots and Systems (IROS), pp. 573–580. IEEE (2012)

498. Su, B., Ma, J., Peng, Y., Sheng, M.: Algorithm for RGBD point cloud denoising and simplification based on k-means clustering **28**, 2329–2334 and 2341 (2016)

499. Suhr, J., Jang, J., Min, D., Jung, H.: Sensor fusion-based low-cost vehicle localization system for complex urban environments. IEEE Trans. Intell. Transport. Syst. **18**, 1–9 (2016). https://doi.org/10.1109/TITS.2016.2595618

500. Sumikura, S., Shibuya, M., Sakurada, K.: OpenVSLAM: a versatile visual SLAM framework, pp. 2292–2295 (2019). https://doi.org/10.1145/3343031.3350539
501. Sun, P., Kretzschmar, H., Dotiwalla, X., Chouard, A., Patnaik, V., Tsui, P., Guo, J., Zhou, Y., Chai, Y., Caine, B., et al.: Scalability in perception for autonomous driving: Waymo open dataset. In: 2020 IEEE/CVF Conference on Computer Vision and Pattern Recognition (CVPR) (2019). arXiv–1912
502. Taghavi, S., Shi, W.: EdgeMask: An edge-based privacy preserving service for video data sharing. In: Proceedings of the 3rd Workshop on Security and Privacy in Edge Computing (EdgeSP) (2020)
503. Takaya, K., Asai, T., Kroumov, V., Smarandache, F.: Simulation environment for mobile robots testing using ros and gazebo. In: 2016 20th International Conference on System Theory, Control and Computing (ICSTCC), pp. 96–101. IEEE, New York (2016)
504. Takleh, T.T.O., Bakar, N.A., Rahman, S.A., Hamzah, R., Aziz, Z.: A brief survey on slam methods in autonomous vehicle. Int. J. Eng. Technol. **7**(4), 38–43 (2018)
505. Talagala, N., Sundararaman, S., Sridhar, V., Arteaga, D., Luo, Q., Subramanian, S., Ghanta, S., Khermosh, L., Roselli, D.: ECO: Harmonizing edge and cloud with ML/DL orchestration. In: USENIX Workshop on Hot Topics in Edge Computing (HotEdge 18) (2018)
506. Tamai, Y., Hasegawa, T., Ozawa, S.: The ego-lane detection under rainy condition. In: World Congress on Intelligent Transport Systems (3rd: 1996: Orlando Fla.). Intelligent Transportation: Realizing the Future: Abstracts of the Third World Congress on Intelligent Transport Systems (1996)
507. Tampuu, A., Semikin, M., Muhammad, N., Fishman, D., Matiisen, T.: A survey of end-to-end driving: architectures and training methods (2020). Preprint. arXiv:2003.06404
508. Tan, H., Zhou, Y., Zhu, Y., Yao, D., Li, K.: A novel curve lane detection based on improved river flow and RANSA. In: 17th International IEEE Conference on Intelligent Transportation Systems (ITSC), pp. 133–138. IEEE, New York (2014)
509. Tang, L., Shi, Y., He, Q., Sadek, A.W., Qiao, C.: Performance test of autonomous vehicle LiDAR sensors under different weather conditions. Transport. Res. Record **2674**(1), 319–329 (2020)
510. Teichman, A., Levinson, J., Thrun, S.: Towards 3D object recognition via classification of arbitrary object tracks. In: 2011 IEEE International Conference on Robotics and Automation, pp. 4034–4041. IEEE, New York (2011)
511. Teng, Z., Kim, J.H., Kang, D.J.: Real-time lane detection by using multiple cues. In: ICCAS 2010, pp. 2334–2337. IEEE, New York (2010)
512. Tesla: Tesla Autopilot: Full Self-Driving Hardware on All Cars (2018). https://www.tesla.com/autopilot
513. Texas Instruments TDA. http://www.ti.com/processors/automotive-processors/tdax-adas-socs/overview.html. Accessed 28 Dec 2018
514. The basics of LiDAR - light detection and ranging - remote sensing (2020). https://www.neonscience.org/lidar-basics. Accessed 18 Feb 2020
515. The world's first self-driving ubers are on the road in the steel city (2017). https://www.uber.com/cities/pittsburgh/self-driving-ubers/
516. The CAR Lab (2020). https://www.thecarlab.org/
517. The Evolution of EyeQ (2020). https://www.mobileye.com/our-technology/evolution-eyeq-chip/
518. The challenges of developing autonomous vehicles during a pandemic. http://bit.ly/ChallengesAD (2020). Online. Accessed 01 Dec 2020
519. Thrun, S., Leonard, J.J.: Simultaneous localization and mapping. In: Springer Handbook of Robotics, pp. 871–889. Springer, New York (2008)
520. Tian, Y., Pei, K., Jana, S., Ray, B.: Deeptest: Automated testing of deep-neural-network-driven autonomous cars. In: Proceedings of the 40th International Conference on Software Engineering, pp. 303–314 (2018)

521. Tijtgat, N., Van Ranst, W., Goedeme, T., Volckaert, B., De Turck, F.: Embedded real-time object detection for a UAV warning system. In: Proceedings of the IEEE International Conference on Computer Vision, pp. 2110–2118 (2017)

522. Toderici, G., Vincent, D., Johnston, N., Jin Hwang, S., Minnen, D., Shor, J., Covell, M.: Full resolution image compression with recurrent neural networks. In: The IEEE Conference on Computer Vision and Pattern Recognition (CVPR) (2017)

523. Tomic, T., Schmid, K., Lutz, P., Domel, A., Kassecker, M., Mair, E., Grixa, I.L., Ruess, F., Suppa, M., Burschka, D.: Toward a fully autonomous UAV: Research platform for indoor and outdoor urban search and rescue. IEEE Robot. Autom. Mag. **19**(3), 46–56 (2012)

524. Toulminet, G., Boussuge, J., Laurgeau, C.: Comparative synthesis of the 3 main European projects dealing with cooperative systems (CVIS, SAFESPOT and COOPERS) and description of COOPERS demonstration site 4. In: 11th International IEEE Conference on Intelligent Transportation Systems, 2008. ITSC 2008, pp. 809–814. IEEE, New York (2008)

525. Tran, T.X., Chan, K., Pompili, D.: COSTA: Cost-aware service caching and task offloading assignment in mobile-edge computing. In: 2019 16th Annual IEEE International Conference on Sensing, Communication, and Networking (SECON), pp. 1–9. IEEE, New York (2019)

526. tusimple: tusimple-benchmark. [EB/OL]. https://github.com/TuSimple/tusimple-benchmark/. Accessed 4 Jan 2021

527. udacity: udacity. [EB/OL]. https://github.com/udacity/self-driving-car/. Accessed 4 Jan 2021

528. Ulm, G., Gustavsson, E., Jirstrand, M.: OODIDA: on-board/off-board distributed data analytics for connected vehicles (2019). Preprint. arXiv:1902.00319

529. Unity: Unity technologies (2021). https://unity.com/

530. Unreal: Unreal engine technologies (2021). https://www.unrealengine.com/en-US/

531. Urmson, C., Anhalt, J., Bagnell, D., Baker, C., Bittner, R., Clark, M., Dolan, J., Duggins, D., Galatali, T., Geyer, C., et al.: Autonomous driving in urban environments: Boss and the urban challenge. J. Field Robot. **25**(8), 425–466 (2008)

532. Urmson, C., Anhalt, J., Clark, M., Galatali, T., Gonzalez, J., Gowdy, J., Gutierrez, A., Harbaugh, S., Johnson-Roberson, M., Koon, P., Peterson, K., Smith, B.: High speed navigation of unrehearsed terrain: red team technology for grand challenge 2004. Robotics Institute, Carnegie Mellon University, Pittsburgh, PA, Tech. Rep. CMU-RI-04-37 (2004)

533. USRP B210 (2019). https://www.ettus.com/all-products/ub210-kit/. Available at https://www.ettus.com/all-products/ub210-kit/

534. Van Brummelen, J., O'Brien, M., Gruyer, D., Najjaran, H.: Autonomous vehicle perception: the technology of today and tomorrow. Transport. Res. Part C: Emerg. Technol. **89**, 384–406 (2018)

535. Varma, G., Subramanian, A., Namboodiri, A., Chandraker, M., Jawahar, C.V.: IDD: a dataset for exploring problems of autonomous navigation in unconstrained environments. In: 2019 IEEE Winter Conference on Applications of Computer Vision (WACV) (2018)

536. Velodyne LiDAR products (2019). https://velodynelidar.com/products.html

537. Venkata, S.K., Ahn, I., Jeon, D., Gupta, A., Louie, C., Garcia, S., Belongie, S., Taylor, M.B.: SD-VBS: The San Diego vision benchmark suite. In: 2009 IEEE International Symposium on Workload Characterization (IISWC), pp. 55–64. IEEE, New York (2009)

538. VGG image annotator (2021). https://gitlab.com/vgg/via

539. Viola, P., Jones, M.: Rapid object detection using a boosted cascade of simple features. In: Proceedings of the 2001 IEEE Computer Society Conference on Computer Vision and Pattern Recognition. CVPR 2001, vol. 1, pp. I–I. IEEE, New York (2001)

540. Vulimiri, A., Curino, C., Godfrey, P.B., Jungblut, T., Karanasos, K., Padhye, J., Varghese, G.: Wanalytics: geo-distributed analytics for a data intensive world. In: Proceedings of the 2015 ACM SIGMOD International Conference on Management of Data, pp. 1087–1092. ACM, New York (2015)

541. VxWorks. https://www.windriver.com/products/vxworks/. Accessed 28 Dec 2018

542. Wang, J., Feng, Z., Chen, Z., George, S., Bala, M., Pillai, P., Yang, S.W., Satyanarayanan, M.: Bandwidth-efficient live video analytics for drones via edge computing. In: 2018 IEEE/ACM Symposium on Edge Computing (SEC), pp. 159–173. IEEE, New York (2018)

543. Wang, J., Feng, Z., George, S., Iyengar, R., Pillai, P., Satyanarayanan, M.: Towards scalable edge-native applications. In: Proceedings of the 4th ACM/IEEE Symposium on Edge Computing, pp. 152–165 (2019)

544. Wang, L., Zhan, J., Luo, C., Zhu, Y., Yang, Q., He, Y., Gao, W., Jia, Z., Shi, Y., Zhang, S., et al.: BigDataBench: a big data benchmark suite from internet services. In: 2014 IEEE 20th International Symposium on High Performance Computer Architecture (HPCA), pp. 488–499. IEEE, New York (2014)

545. Wang, R., Azab, A.M., Enck, W., Li, N., Ning, P., Chen, X., Shen, W., Cheng, Y.: Spoke: scalable knowledge collection and attack surface analysis of access control policy for security enhanced android. In: Proceedings of the 2017 ACM on Asia Conference on Computer and Communications Security, ASIA CCS '17, p. 612–624. Association for Computing Machinery, New York (2017). https://doi.org/10.1145/3052973.3052991

546. Wang, X., Ning, Z., Wang, L.: Offloading in internet of vehicles: a fog-enabled real-time traffic management system. IEEE Trans. Ind. Inf. **14**(10), 4568–4578 (2018)

547. Wang, Y., Chao, W.L., Garg, D., Hariharan, B., Campbell, M., Weinberger, K.Q.: Pseudo-lidar from visual depth estimation: bridging the gap in 3d object detection for autonomous driving. In: Proceedings of the IEEE/CVF Conference on Computer Vision and Pattern Recognition (CVPR) (2019)

548. Wang, Y., Liu, L., Zhang, X., Shi, W.: HydraOne: an indoor experimental research and education platform for CAVs. In: 2nd {USENIX} Workshop on Hot Topics in Edge Computing (HotEdge 19) (2019)

549. Wang, Y., Liu, S., Wu, X., Shi, W.: Cavbench: a benchmark suite for connected and autonomous vehicles. In: 2018 IEEE/ACM Symposium on Edge Computing (SEC), pp. 30–42. IEEE, New York (2018)

550. Wang, Y., Weinacker, H., Koch, B.: A LiDAR point cloud based procedure for vertical canopy structure analysis and 3d single tree modelling in forest. Sensors **8**(6), 3938–3951 (2008). http://dx.doi.org/10.3390/s8063938

551. Warrender, C., Forrest, S., Pearlmutter, B.: Detecting intrusions using system calls: alternative data models. In: Proceedings of the 1999 IEEE Symposium on Security and Privacy (Cat. No.99CB36344), pp. 133–145 (1999)

552. Waymo: Waymo Self-Driving Car (2018). https://waymo.com

553. Waymo is using AI to simulate autonomous vehicle camera data. http://bit.ly/WaymoAI (2020). Online. Accessed 01 Dec 2020

554. Waymo (2017). https://waymo.com/

555. Off road, but not offline: How simulation helps advance our waymo driver. http://bit.ly/WaymoBlog (2020). Online. Accessed 01 Dec 2020

556. Wei, J., Snider, J.M., Kim, J., Dolan, J.M., Rajkumar, R., Litkouhi, B.: Towards a viable autonomous driving research platform. In: 2013 IEEE Intelligent Vehicles Symposium (IV), pp. 763–770 (2013). https://doi.org/10.1109/IVS.2013.6629559

557. Wei, J., Snider, J.M., Kim, J., Dolan, J.M., Rajkumar, R., Litkouhi, B.: Towards a viable autonomous driving research platform. In: 2013 IEEE Intelligent Vehicles Symposium (IV), pp. 763–770. IEEE, New York (2013)

558. White, R., Christensen, H.I., Caiazza, G., Cortesi, A.: Procedurally provisioned access control for robotic systems. In: 2018 IEEE/RSJ International Conference on Intelligent Robots and Systems (IROS), pp. 1–9 (2018). https://doi.org/10.1109/IROS.2018.8594462

559. Wilhelm, R., Engblom, J., Ermedahl, A., Holsti, N., Thesing, S., Whalley, D., Bernat, G., Ferdinand, C., Heckmann, R., Mitra, T., et al.: The worst-case execution-time problem—overview of methods and survey of tools. ACM Trans. Embedded Comput. Syst. **7**(3), 1–53 (2008)

560. Wilson, S., Gameros, R., Sheely, M., Lin, M., Dover, K., Gevorkyan, R., Haberland, M., Bertozzi, A., Berman, S.: Pheeno, a versatile swarm robotic research and education platform. IEEE Robot. Autom. Lett. **1**(2), 884–891 (2016)

561. Wojke, N., Bewley, A.: Deep cosine metric learning for person re-identification. In: 2018 IEEE Winter Conference on Applications of Computer Vision (WACV), pp. 748–756. IEEE, New York (2018)

562. Wojke, N., Bewley, A., Paulus, D.: Simple online and realtime tracking with a deep association metric. In: 2017 IEEE International Conference on Image Processing (ICIP), pp. 3645–3649. IEEE, New York (2017)

563. Wolcott, R., Eustice, R.: Visual localization within LiDAR maps for automated urban driving. In: IEEE International Conference on Intelligent Robots and Systems, pp. 176–183 (2014). https://doi.org/10.1109/IROS.2014.6942558

564. Wu, C.J., Brooks, D., Chen, K., Chen, D., Choudhury, S., Dukhan, M., Hazelwood, K., Isaac, E., Jia, Y., Jia, B., et al.: Machine learning at facebook: understanding inference at the edge. In: 2019 IEEE International Symposium on High Performance Computer Architecture (HPCA), pp. 331–344. IEEE, New York (2019)

565. Wu, P.C., Chang, C.Y., Lin, C.H.: Lane-mark extraction for automobiles under complex conditions. Patt. Recogn. **47**(8), 2756–2767 (2014)

566. Xiang, Y., Choi, W., Lin, Y., Savarese, S.: Subcategory-aware convolutional neural networks for object proposals and detection. In: 2017 IEEE Winter Conference on Applications of Computer Vision (WACV) (2017)

567. Xiang, Y., Mottaghi, R., Savarese, S.: Beyond PASCAL: A benchmark for 3D object detection in the wild. In: 2014 IEEE Winter Conference on Applications of Computer Vision (WACV), pp. 75–82. IEEE, New York (2014)

568. Xie, X., Yu, Y., Lin, X., Sun, C.: An EKF SLAM algorithm for mobile robot with sensor bias estimation. In: 2017 32nd Youth Academic Annual Conference of Chinese Association of Automation (YAC), pp. 281–285. IEEE, New York (2017)

569. Xing, Y., Ma, E.W., Tsui, K.L., Pecht, M.: Battery management systems in electric and hybrid vehicles. Energies **4**(11), 1840–1857 (2011)

570. Xiong, Z., Li, W., Han, Q., Cai, Z.: Privacy-preserving auto-driving: a GAN-based approach to protect vehicular camera data. In: 2019 IEEE International Conference on Data Mining (ICDM), pp. 668–677 (2019). https://doi.org/10.1109/ICDM.2019.00077

571. Xu, H., Gao, Y., Yu, F., Darrell, T.: End-to-end learning of driving models from large-scale video datasets. In: Proceedings of the IEEE Conference on Computer Vision and Pattern Recognition, pp. 2174–2182 (2017)

572. Xu, Z., Peng, X., Zhang, L., Li, D., Sun, N.: The ϕ-stack for smart web of things. In: Proceedings of the Workshop on Smart Internet of Things, p. 10. ACM, New York (2017)

573. Yamazaki, S., Miyajima, C., Yurtsever, E., Takeda, K., Mori, M., Hitomi, K., Egawa, M.: Integrating driving behavior and traffic context through signal symbolization. In: 2016 IEEE Intelligent Vehicles Symposium (IV), pp. 642–647. IEEE, New York (2016)

574. Yan, C., Xu, W., Liu, J.: Can you trust autonomous vehicles: contactless attacks against sensors of self-driving vehicle, p. 109 (2016). Corpus ID: 27264520

575. Yan, X.W., Guo, Y.W., Cui, Y., Wang, Y.W., Deng, H.R.: Electric vehicle battery SOC estimation based on GNL model adaptive Kalman filter. J. Phys. Conf. Ser. **1087**, 052027. IOP Publishing (2018)

576. Yang, S., Song, Y., Kaess, M., Scherer, S.: Pop-up SLAM: semantic monocular plane SLAM for low-texture environments. pp. 1222–1229 (2016). https://doi.org/10.1109/IROS.2016.7759204

577. Yang, T.J., Chen, Y.H., Sze, V.: Designing energy-efficient convolutional neural networks using energy-aware pruning. In: Proceedings of the IEEE Conference on Computer Vision and Pattern Recognition, pp. 5687–5695 (2017)

578. Yang, Z., Sun, Y., Liu, S., Shen, X., Jia, J.: Std: Sparse-to-dense 3d object detector for point cloud. In: Proceedings of the IEEE International Conference on Computer Vision, pp. 1951–1960 (2019)

579. Yao, W., Dai, W., Xiao, J., Lu, H., Zheng, Z.: A simulation system based on ros and gazebo for robocup middle size league. In: 2015 IEEE International Conference on Robotics and Biomimetics (ROBIO), pp. 54–59. IEEE, New York (2015)

580. Yağdereli, E., Gemci, C., Aktaş, A.Z.: A study on cyber-security of autonomous and unmanned vehicles. J. Defense Model. Simul. **12**(4), 369–381 (2015). https://doi.org/10.1177/1548512915575803

581. Ye, J., Chow, J.H., Chen, J., Zheng, Z.: Stochastic gradient boosted distributed decision trees. In: Proceedings of the 18th ACM Conference on Information and Knowledge Management, pp. 2061–2064. ACM, New York (2009)

582. Yi, S., Hao, Z., Zhang, Q., Zhang, Q., Shi, W., Li, Q.: LAVEA: latency-aware video analytics on edge computing platform. In: Proceedings of the 2nd ACM/IEEE Symposium on Edge Computing, pp. 1–13 (2017)

583. Yin, H., Berger, C.: When to use what data set for your self-driving car algorithm: An overview of publicly available driving datasets. In: 2017 IEEE 20th International Conference on Intelligent Transportation Systems (ITSC), pp. 1–8. IEEE, New York (2017)

584. Yodaiken, V., et al.: The RTLinux manifesto. In: Proceedings of the 5th Linux Expo (1999)

585. You, C.W., Lane, N.D., Chen, F., Wang, R., Chen, Z., Bao, T.J., Montes-de Oca, M., Cheng, Y., Lin, M., Torresani, L., et al.: Carsafe app: alerting drowsy and distracted drivers using dual cameras on smartphones. In: Proceeding of the 11th Annual International Conference on Mobile Systems, Applications, and Services, pp. 13–26. ACM, New York (2013)

586. Yu, F., Xian, W., Chen, Y., Liu, F., Liao, M., Madhavan, V., Darrell, T.: BDD100K: a diverse driving video database with scalable annotation tooling. 2(5), 6 (2018). Preprint. arXiv:1805.04687

587. Yu, J., guo, x., Pei, X., chen, z., Zhu, M.: Robust model predictive control for path tracking of autonomous vehicle (2019). https://doi.org/10.4271/2019-01-0693

588. Yuan, J., Zheng, Y., Xie, X., Sun, G.: Driving with knowledge from the physical world. In: Proceedings of the 17th ACM SIGKDD International Conference on Knowledge Discovery and Data Mining, pp. 316–324. ACM, New York (2011)

589. Yurtsever, E., Lambert, J., Carballo, A., Takeda, K.: A survey of autonomous driving: common practices and emerging technologies. IEEE Access 8, 58443–58469 (2020)

590. ZainEldin, H., Elhosseini, M.A., Ali, H.A.: Image compression algorithms in wireless multimedia sensor networks: a survey. Ain Shams Eng. J. 6(2), 481–490 (2015). https://doi.org/10.1016/j.asej.2014.11.001. http://www.sciencedirect.com/science/article/pii/S2090447914001567

591. Zang, S., Ding, M., Smith, D., Tyler, P., Rakotoarivelo, T., Kaafar, M.A.: The impact of adverse weather conditions on autonomous vehicles: How rain, snow, fog, and hail affect the performance of a self-driving car. IEEE Veh. Technol. Mag. 14(2), 103–111 (2019)

592. Zeng, K.C., Liu, S., Shu, Y., Wang, D., Li, H., Dou, Y., Wang, G., Yang, Y.: All your GPS are belong to us: towards stealthy manipulation of road navigation systems. In: 27th USENIX Security Symposium (USENIX Security 18), pp. 1527–1544. USENIX Association, Baltimore, MD (2018). https://www.usenix.org/conference/usenixsecurity18/presentation/zeng

593. Zhang, D., Ma, Y., Zheng, C., Zhang, Y., Hu, X.S., Wang, D.: Cooperative-competitive task allocation in edge computing for delay-sensitive social sensing. In: 2018 IEEE/ACM Symposium on Edge Computing (SEC), pp. 243–259. IEEE, New York (2018)

594. Zhang, F., Guan, C., Fang, J., Bai, S., Yang, R., Torr, P.H., Prisacariu, V.: Instance segmentation of lidar point clouds. In: 2020 IEEE International Conference on Robotics and Automation (ICRA), pp. 9448–9455. IEEE, New York (2020)

595. Zhang, F., Li, S., Yuan, S., Sun, E., Zhao, L.: Algorithms analysis of mobile robot slam based on kalman and particle filter. In: 2017 9th International Conference on Modelling, Identification and Control (ICMIC), pp. 1050–1055. IEEE, New York (2017)

596. Zhang, G., Yan, C., Ji, X., Zhang, T., Zhang, T., Xu, W.: DolphinAttack: inaudible voice commands. In: Proceedings of the 2017 ACM SIGSAC Conference on Computer and Communications Security, CCS '17, p. 103–117. Association for Computing Machinery, New York (2017). https://doi.org/10.1145/3133956.3134052

597. Zhang, J., Lee, J.: A review on prognostics and health monitoring of li-ion battery. J. Power Sources 196(15), 6007–6014 (2011)

598. Zhang, J., Singh, S.: LOAM: lidar odometry and mapping in real-time (2014). https://doi.org/10.15607/RSS.2014.X.007

599. Zhang, P., Zhou, M., Fortino, G.: Security and trust issues in fog computing: a survey. Fut. Gener. Comput. Syst. **88**, 16–27 (2018)

600. Zhang, Q., Wang, Y., Zhang, X., Liu, L., Wu, X., Shi, W., Zhong, H.: OpenVDAP: an open vehicular data analytics platform for CAVs. In: 2018 IEEE 38th International Conference on Distributed Computing Systems (ICDCS), pp. 1310–1320. IEEE, New York (2018)

601. Zhang, Q., Zhang, Q., Shi, W., Zhong, H.: Enhancing AMBER alert using collaborative edges: poster. In: Proceedings of the Second ACM/IEEE Symposium on Edge Computing, p. 27. ACM, New York (2017)

602. Zhang, Q., Zhang, Q., Shi, W., Zhong, H.: Firework: Data processing and sharing for hybrid cloud-edge analytics. Technical Report MIST-TR-2017-002 (2017)

603. Zhang, Q., Zhang, Q., Shi, W., Zhong, H.: Distributed collaborative execution on the edges and its application to amber alerts. IEEE Internet Things J. **5**(5), 3580–3593 (2018). https://doi.org/10.1109/JIOT.2018.2845898

604. Zhang, Q., Zhong, H., Cui, J., Ren, L., Shi, W.: AC4AV: a flexible and dynamic access control framework for connected and autonomous vehicles. IEEE Internet Things J. **PP**, 1–1 (2020)

605. Zhang, Y., Gantt, G.W., Rychlinski, M.J., Edwards, R.M., Correia, J.J., Wolf, C.E.: Connected vehicle diagnostics and prognostics, concept, and initial practice. IEEE Trans. Reliab. **58**(2), 286–294 (2009)

606. Zhang, Z., Liu, S., Tsai, G., Hu, H., Chu, C.C., Zheng, F.: PIRVS: an advanced visual-inertial SLAM system with flexible sensor fusion and hardware co-design. In: 2018 IEEE International Conference on Robotics and Automation (ICRA), pp. 1–7. IEEE, New York (2018)

607. Zheng, K., Zheng, Q., Chatzimisios, P., Xiang, W., Zhou, Y.: Heterogeneous vehicular networking: a survey on architecture, challenges, and solutions. IEEE Commun. Surv. Tutor. **17**(4), 2377–2396 (2015)

608. Zheng, K., Zheng, Q., Yang, H., Zhao, L., Hou, L., Chatzimisios, P.: Reliable and efficient autonomous driving: the need for heterogeneous vehicular networks. IEEE Commun. Mag. **53**(12), 72–79 (2015)

609. Zhong, H., Pan, L., Zhang, Q., Cui, J.: A new message authentication scheme for multiple devices in intelligent connected vehicles based on edge computing. IEEE Access **7**, 108211–108222 (2019)

610. Zhou, N., Li, H., Wang, D., Pan, S., Zhou, Z.: Image compression and encryption scheme based on 2d compressive sensing and fractional mellin transform. Opt. Commun. **343**, 10–21 (2015). https://doi.org/10.1016/j.optcom.2014.12.084. http://www.sciencedirect.com/science/article/pii/S0030401815000048

611. Zhou, S., Jiang, Y., Xi, J., Gong, J., Xiong, G., Chen, H.: A novel lane detection based on geometrical model and gabor filter. In: 2010 IEEE Intelligent Vehicles Symposium, pp. 59–64. IEEE, New York (2010)

612. Zhou, Y., Wang, C., Zhou, X.: DCT-based color image compression algorithm using an efficient lossless encoder. In: 2018 14th IEEE International Conference on Signal Processing (ICSP), pp. 450–454 (2018)

613. Zhu, M., Gupta, S.: To prune, or not to prune: exploring the efficacy of pruning for model compression (2017). Preprint. arXiv:1710.01878

614. Zhu, Z., Liang, D., Zhang, S., Huang, X., Li, B., Hu, S.: Traffic-sign detection and classification in the wild. In: Proceedings of the IEEE Conference on Computer Vision and Pattern Recognition, pp. 2110–2118 (2016)

615. Ziegler, J., Bender, P., Schreiber, M., Lategahn, H., Strauss, T., Stiller, C., Dang, T., Franke, U., Appenrodt, N., Keller, C.G., et al.: Making bertha drive—an autonomous journey on a historic route. IEEE Intell. Transport. Syst. Mag. **6**(2), 8–20 (2014)

616. Zolanvari, I., Ruano, S., Rana, A., Cummins, A., Smolic, A., Da Silva, R., Rahbar, M.: DublinCity: annotated LiDAR point cloud and its applications (2019)

617. Zou, Z., Shi, Z., Guo, Y., Ye, J.: Object detection in 20 years: a survey (2019). Preprint. arXiv:1905.05055

618. Zynq UltraScale+ MPSoC ZCU104 Evaluation Kit (2020). https://www.xilinx.com/products/boards-and-kits/zcu104.html

Index

Printed in the United States
by Baker & Taylor Publisher Services